Higher-Order Components for Grid Programming

Jan Dünnweber · Sergei Gorlatch

Higher-Order Components for Grid Programming

Making Grids More Usable

 Springer

Jan Dünnweber
Senior IT-Consultant
beck et al. projects GmbH
Theresienhöhe 13
80339 München
Germany
Jan.Duennweber@bea.de

Sergei Gorlatch
Universität Münster
Institut für Informatik
Einsteinstr. 62
48149 Münster
Germany
gorlatch@math.uni-muenster.de

ISBN 978-3-642-42501-1   ISBN 978-3-642-00841-2 (eBook)
DOI 10.1007/978-3-642-00841-2
Springer Dordrecht Heidelberg London New York

ACM Computing Classification (1998): C.2.4, D.2.11, D.2.12, H.3.5, J.2

© Springer-Verlag Berlin Heidelberg 2009
Softcover re-print of the Hardcover 1st edition 2009

Cover design: KünkelLopka, Heidelberg

Printed on acid-free paper

Springer is part of Springer Science+Business Media (www.springer.com)

Preface

I Abstract

Grid computing is an emerging technology that allows the users to conveniently access and share different kinds of resources – processing, data and networks – which are distributed worldwide over the Internet. Grids promise for broad user communities a transparent access to these resources, similar to today's user-friendly access to electricity by plugging into the power grid. Grid technology is viewed as an enabler for both high-performance and data-intensive computing, and is therefore sometimes called the future Internet. The application fields of grid computing are manyfold and include high-energy physics, crash test simulations, climate research, computational biology, etc.

A major challenge in grid computing remains the application software development for this new kind of infrastructure. Grid application programmers have to take into account several complicated aspects: distribution of data and computations, parallel computations between different sites and processors, heterogeneity of the involved computers, load balancing, etc. This makes grid programming a cumbersome task which might inhibit the acceptance of grids by a broad users community. The contemporary grid middleware systems – Globus and others – tackle some of these problems; however, they require from the programmer to explicitly provide a specific middleware setup using special formats and tools like WSDL and others. This middleware-specific setup must be typically done from scratch for every new application and completely rewritten if the application is changing, thus making the programmer's task even more complex and reducing the software re-use. It is widely agreed that new programming models for grids are necessary which hide from the application software developer the details of the underlying grid infrastructure and middleware.

This book presents *Higher-Order Components (HOCs)* – a novel grid programming model, and puts it into a broader context of the contemporary distributed programming. HOCs are generic software components that can be parameterized not only with data but also with application-specific code (thus, the name higher-order). HOCs implement various patterns of parallel and distributed processing, such as farm, pipeline, divide-and-conquer and others. Implementations of components are provided to the programmer via a high-level, Internet-accessible service interface. A HOC implementation includes all necessary parallelization and synchronization and also the required middleware setup, which remain hidden from the programmer. The programmer develops an application by composing suitable HOCs and customizing them for the particular use case by means of application-specific parameters (which may be either data or code). Thus, the application developer is concerned with the algorithm and other relevant aspects of the particular application, while the low-level concerns of the required middleware setup and the internal communication and synchronization within the employed parallel patterns remain hidden from him as a part of the HOC implementation.

The book describes the specific HOCs' features within the contemporary landscape of the component-based approaches in general and grid components in particular. We pay special attention to the Fractal component model which serves as a basis for the unified European Grid Component Model (GCM) which has been developed within the CoreGRID Network of Excellence. Both models – Fractal and GCM – provide a rich, well-founded set of composition opportunities which enable an efficient component-based development of complex applications. We describe how our higher-order components add to these features the possibility to use high-performance implementations of parallel processing patterns, and to employ standard Web services for exchanging data and code between HOCs and other application entities. The choice of Web services as the underlying communication technology makes HOCs interoperable with Fractal, GCM, and other contemporary distributed technologies.

In addition to general considerations, we present in detail the implementation approach taken for higher-order components. The book describes HOC-SA – the service architecture for HOCs – and its use as the runtime-system for component-based grid applications. We explain the relation between HOC-SA and the very popular Globus middleware. Due to its novel code transfer mechanism, HOC-SA was extensively evaluated and then adopted as an optional extension to the Globus toolkit as so-called Globus incubator project, available from the Globus site: www.globus.org.

To demonstrate the usefulness of the HOC approach and to provide the reader with the hands-on experience, we describe a broad collection of example applications from various fields of science and engineering, including biology, physics, etc. The Java code for these examples is provided online, complementary to the book. All examples use the HOC Service Architecture (HOC-SA) as their runtime environment. The expected application performance is studied and reported for extensive performance experiments on different testbeds, including grids with worldwide distribution.

The book is targeted at graduate students, software developers, and researchers in both academia and industry. Readers can raise their level of knowledge about methodologies for programming contemporary parallel and distributed systems, and, furthermore, gain practical experience in using distributed software, by picking out topics from the presented material. Practical examples show how the complementary online material can easily be adopted into various new projects.

II Structure of this Book

The book is organized in seven chapters:

Chapter 1 provides an introduction into the area of modern distributed systems and application programming for such systems. We outline the main contemporary approaches to distributed programming and show the benefits and implications of using distributed middleware. In particular, we cover the most popular middleware systems: RMI and CORBA, as well as Web-enabled systems like Tomcat and Globus. We demonstrate that most distributed applications require the advanced functionalities of a software called a "container" like Globus (which provides HTTP data transmission, persistence, security, etc.) and a simple but efficient communication mechanism. This motivates combinations of multiple technologies, e.g., programs that communicate using RMI and are made accessible to clients via a Web service container. Several examples are used to illustrate how such combinations can be set up.

Chapter 2 introduces the main contribution of this book, Higher-Order Components (HOCs). HOCs represent a generic strategy for the implementation of parallel and distributed computation patterns; these patterns accept executable codes and data as parameters which are supplied to them via the network. By enabling type polymorphism for both program and data parameters, HOCs provide clear benefits in the compositional aspects, as shown by the following two case studies. As the introductory case study, a simple but computation-intensive application is used: the calculation of fractal images which are popular in applications of the so-called chaos theory. Two implementations of this case study, both using multiple distributed servers, are shown: one without components and one using HOCs. The comparison of the two solutions demonstrates the benefits of HOCs for the application programmer. The second case study demonstrates the power of code parameters by showing that HOCs can easily be adapted with user-defined code to different computation schemata and thereby fit into a much broader range of applications than are covered by the traditional component technologies.

Chapter 3 is about Web services, a popular Internet-based communication technology, which is used in this book for allowing remote clients to access program components (HOCs). Besides important advantages (e.g., interoperability among software written in different programming languages), some problems of Web service-based systems are discussed and illustrated by an application of the discrete wavelet transform (DWT), in particular: difficulties in transferring executable code and connecting to traditional high-performance computing technology, such as MPI. For resolving such difficulties, this chapter introduces the HOC Service Architecture (HOC-SA) as a runtime environment for HOCs which extends the range of possible parameter types for any Web service. We discuss the management and configuration issues that arise when executable code is used as a Web service parameter and when code is shared among distributed software components. As a new example, this chapter explains how multiple HOCs can be combined in a single application that adheres to the popular Map-Reduce pattern of parallelism. Finally, we describe the HOC-SA portal – an integrated development environment (IDE) for HOCs – and discuss where the HOC-SA ranges among the other recent developments, such as service-oriented architectures (SOA), the service component architecture (SCA) and the Globus Resource Allocation Manager (WS-GRAM).

Chapter 4 describes applications of HOCs in two different fields, namely, bioinformatics (genome similarity detection) and a physical simulation of deforming objects (e.g., for a car crash test simulation). While the previous examples in this book only operated on synthetically generated input and were used to demonstrate the basic features of HOCs, the applications shown in this chapter are real-world grid applications. The bioinformatics application operates on multiple large genome databases, which could never be fully scanned using a contemporary desktop PC. By the use of a HOC for scanning these databases on the grid with several servers, genome similarities were found which had been previously unknown. The second application allows multiple users to cooperate on the construction of the objects used in the simulation in real time, and thus demonstrates how HOCs can satisfy the requirements of both, multi-user interactive real-time applications and high-performance simulation.

Chapter 5 covers advanced topics, such as the automatic transformation of sequential problem descriptions into parallel code for the execution on top of a distributed platform and the scheduling of component-based, distributed applications on the grid. An analytical cost model for scheduling distributed applications is introduced and experimentally evaluated. An equation system solver is used as an additional case study showing how to apply the previously explained functionalities of software components (i.e., automatic parallelization and scheduling) in concrete applications. It is shown how (and under what circumstances) software development tools from the popular *Apache* Web site, Fractal components and the ProActive grid programming library can be used for creating Web services in grid applications automatically, thus freeing the programmer from dealing with the Web service configuration. Related work, like SOFA components, are also discussed in this chapter.

Chapter 6 ends the book with a concluding discussion of the HOC approach to programming performance-critical applications for distributed platforms and related work. It is shown by examples that the different technologies that recently emerged from Internet-related research areas, namely, the Semantic Web, the Web 2.0 and the grid computing, are complementary to each other: the ontologic classification of data and code in the Semantic Web enables new tactics in algorithms for automatic decision making, which considerably enhance tools like schedulers and code transformation programs (e.g., the ones from the previous chapter); the improved responsiveness of Web 2.0 applications make portal software (such as the HOC-SA portal) much more usable than traditional Web server software, and, thereby, help leveraging grid technologies and tools (e.g., Globus GRAM). An outlook on future work describes planned extensions to the software presented in this book (in particular, for the HOC-SA, which is available open-source for downloading from the Internet). In particular, extensions to the analytical scheduling model from the previous chapter are discussed, which allow users to predict the performance of distributed computations. Such predictions will help in estimating the runtime of application workflows and the implied resource reservation times, which can be useful for billing and accounting for the usage of network resources.

III Acknowledgments

This book gained a lot from the advice, support and comments from many people.

We appreciate the thorough checking of our manuscript by Herbert Kuchen at the University of Münster, the anonymous referees at Springer-Verlag, and our editor, Ralf Gerstner, at Springer-Verlag.

We owe Cătălin Dumitrescu and his colleague Ioan Raicu at the University of Chicago special thanks for their work on the HOC scheduling experiments on PlanetLab and the DAS-2 Platform. Ian Foster's group in Chicago (especially Borja Sotomayor and Jennifer Schopf) supported us in making HOCs evolve into an Incubator Project in the Globus Toolkit.

Within the scope of the CoreGRID Network of Excellence (funded by the European Commission under the Sixth Framework Programme, Project no. FP6-004265) we had fruitful collaborations and discussions with Jens Müller-Iden, Martin Alt, Christian Lengauer, Eduardo Argollo, Martin Griebl, Michael and Philipp Classen, Marco Aldinucci, Sonia Campa, Marco Danelutto, Murray Cole, Anne Benoit, Nikos Parlavantzas, Françoise Baude, Virginie Legrand, Denis Caromel, and Ludovic Henrio.

And last, but not least, we would like to thank our former Diploma students at the University of Münster, Philipp Lüdeking and Johannes Tomasoni, for their dedicated implementation work on the Alignment-HOC and the LooPo-HOC, respectively.

Jan Dünnweber, 28.01.2009, München   Sergei Gorlatch, 28.01.2009, Münster

Contents

Chapter 1
Introduction

The grid harnesses the computational power of multiple heterogeneous computers, interconnected via the Internet. To satisfy the demands of compute-intensive applications, the grid nodes comprise computers of various architectures, such as high-performance servers with shared memory, distributed-memory clusters and ordinary desktop PCs. Sometimes, the grid is referred to as a new computing infrastructure [aFK98] or the Internet of the future.

Applications which are supposed to run on the grid often have processing demands far beyond the capabilities of traditional workstations, mainframes and even supercomputers. Among the most notable examples is the Large Hadron Collider (LHC [cHC07]) at CERN: the physics experiments conducted using LHC produce more than 15 petabytes of data each year. Thus, besides computing power, huge storage resources are required. Applications with similarly high demands stem from bioinformatics, climate research and astronomy.

Programming grid applications is already challenging when applications are "embarrassingly parallel," i.e., without dependencies. Since most "real-world" applications have dependencies, i.e., multiple stages that must be executed in a particular order, using the grid to its full capacity requires, besides the distribution of data, a careful coordination of the distributed activities. An ordinary piece of software, consisting of one monolithic program that handles all the data encoding, distribution and evaluation will quickly become unmaintainable by grid programmers. Therefore, grid applications typically employ middleware, job schedulers, monitoring systems and additional multi-purpose tools. Grid application programmers have to be experts in both, their application domain, and the tools they employ for managing the requirements of the grid platform.

This book presents a novel component-based programming methodology for grid computing. This methodology is based on a component model called *Higher-Order Components* (HOCs) and the *HOC-Service Architecture* (HOC-SA), a runtime environment for HOCs. The HOC-SA enables the grid-wide sharing of code and data and allows Higher-Order Components to accept besides data, executable code in parameters which programmers supply to the HOCs over the Internet. Several case studies described in the book demonstrate that HOCs simplify the programming of

J. Dünnweber, S. Gorlatch, *Higher-Order Components for Grid Programming*,
DOI 10.1007/978-3-642-00841-2_1, © Springer-Verlag Berlin Heidelberg 2009

grid applications considerably and do not imply a serious loss of performance. We show that HOCs abstract over the middleware and allow application programmers to benefit from the presence of multiple tools on the grid without having to control all these tools directly. When using HOCs, programmers rather focus on the logic of their applications and the employed algorithms.

Since component technologies rely on middleware, this introductory chapter about grid middleware is technically fundamental for the following chapters. First, the features of different types of middleware and their advantages for application programmers are shown. Then, the required functionalities for grid applications are identified, which are neither provided by current technologies nor by a combination of them. Without a specific middleware for grid applications (or extensions to the existing technologies), these functionalities thus must fully be handled by the application programmers themselves.

Section 1.1 summarizes the challenges of grid programming and Section 1.2 explains the role of middleware in handling these challenges. Section 1.3 introduces RMI, CORBA and containers for components and services. Here, it is also discussed to what extend servlets and EJBs, which are very popular for Web application programming, can simplify grid application development as well. Section 1.4 summarizes the features that even the most modern middleware, e.g., the Globus toolkit, still lacks of, especially concerning its ease of use for grid application programmers.

1.1 Challenges of Grid Programming

The problem of programming modern Internet-based systems in general and grid systems in particular has its origin in the traditional area of parallel programming which has been actively studied since the 1970s. There are two reasons why parallel programming experience is relevant for the grid. On the one hand, the grid often includes several parallel machines which must be programmed using a suitable parallel programming model. On the other hand, nodes of the grid, either sequential or parallel computers, work simultaneously and in coordination, thus exemplifying various patterns of parallel behavior and interaction.

Despite the progress in parallel programming, computers with many processors are still hard to program. The most popular programming model for parallelism on distributed-memory machines – Message Passing Interface (MPI) – is quite low-level: it requires from the application programmer to provide an explicit, detailed description of all inter-processor communications [bGo04]. Therefore, the programming process is complex and error-prone, especially for emerging massively parallel systems with hundreds and thousands of processors, which retards the progress in many areas with time-intensive applications.

With the introduction of the grid, the difficulties of parallel programming have not disappeared: parallel machines constituting the grid must still be explicitly programmed. Furthermore, the grid causes an additional layer of complexity in the programming process, due to the following reasons:

- Heterogeneity of the compute resources: the machines of the grid may have different architectures and different performance, and they may be available to the user in different configuration at different time.
- Heterogeneity of the network resources: there are large differences in both bandwidth and latency of grid interconnects over the system space and over time.
- Complex scheduling: since the different parts of any non-trivial grid application depend on each other, a non-optimal assignment of jobs to the servers may lead to big delays in the overall application processing.

The current answer to the specific challenge of complexity in the grid programming is *grid middleware*, whose purpose, reflected in its name, is to mediate between the system and the application programmer. It can hardly be expected from the application programmer to take care of multiple, heterogeneous and highly dynamic resources of a grid system when designing an application. First, the human programmer is simply overburdened by the sheer scale of the system and its highly dynamic nature. Second, the quality requirements to grid programs are very high since numerous and expensive resources are involved. Therefore, grid programming challenges are assumed to be met by the grid middleware which plays a central role in the current realm of grid software.

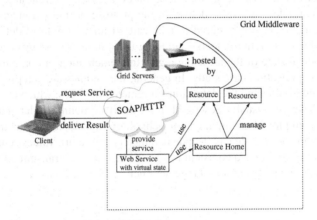

Figure 1.1 Grid Programming Using Middleware

Figure 1.1 shows the scenario of using middleware on the grid via Web services and resources as envisaged by the currently very popular Web Service Resource Framework middleware architecture (WSRF [cOA04]). An application located on the client issues a request using SOAP via HTTP (the standard application-layer and transport protocols for Web service communication [cW302]) for a particular service (i.e., a program to be executed remotely) required from the grid. When replying to the user's request, the grid middleware decides on where the required service should come from. The middleware also enables the user to access so-called *resource homes* and WS *resources* which are hosted by the grid servers; in the WSRF model, resources are not hardware but data records used for representing the server-side data stock. Without such WS resources, data could not be shared among the distributed servers and the client so efficiently, since the full application state data had to be transferred in every single Web service communication. The resource homes manage the resources, allowing clients to request the creation of new resources over the network. Using the remote resources, clients can maintain a distributed service state, called *virtual state* in Fig. 1.1. The service state is virtual, in the sense that the state data does not belong to the service, but the client must hold up the connection between each service and the corresponding state data by specifying the affected resources when sending service requests. Note that some texts call Web services which are connected to WS resources *grid services*. This book avoids the term grid services, since it is ambiguous: In the OGSI specification [cGl96], which is outdated now, a grid service was something completely different.

Once a service in the WSRF completes, the computed result is sent back to the user. From the high-level point of view, the user obtains the requested service transparently, which can be expressed as follows: the user requests – the system delivers. In this ideal scenario, the application programmer is shielded from the complexity of grid programming and is expected to concentrate on the really important aspects of the application: the algorithm, its accuracy, the aspects of performance, etc.

In contrast to this ideal scenario, grid middleware, in its current and short-term anticipated state, does not fully deliver to the promise of transparency: the application programmer is still required to take care of many low-level details. Chapter 2 illustrates these requirements (including the middleware setup) according to the technological state of the art using a grid implementation of computing fractal images as an application case study. In general, the middleware setup includes the resource configuration and the definitions of remote interfaces, which are not trivial in the grid. They are usually described in an interface definition language (IDL), specific to the employed communication technology, in order to instruct the runtime environment how to encode and decode parameters and return values sent over network boundaries (e.g., via a conversion from XML elements transmitted via SOAP into primitives of a programming language, such as Java, and vice versa).

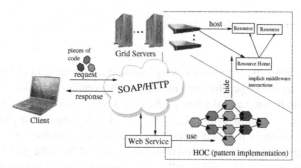

Figure 1.2 Grid Programming Using HOCs

Figure 1.2 shows how the ideal scenario is approached, when a HOC is used, as proposed in this book. The clients still connect to the grid via a Web service. But now this service belongs to a HOC expressing a well-defined pattern of parallel processing which is made accessible over the Internet via the Web service. To customize the HOC for the application, the client transfers pieces of application-specific code as parameters which are executed within the pattern implementation. All the middleware interactions (network communication and resource management) are contained in the HOC implementation which hides them, as well as the required middleware setup for running the HOC, from the user.

The Web services running on top of the Globus middleware employ Internet protocols (such as HTTP as shown in Fig. 1.1 and Fig. 1.2) for communication and an XML-based IDL (WSDL [cW306]) for service interface definitions. This helps leveraging service-oriented architectures (SOAs [aEr04]) which are ideal for running distributed computations on heterogeneous and dynamic platforms like the grid, since the connections among physical machines can easily be exchanged (which is often referred to as the principle of *loose coupling* [aKa03]). Modern grid middleware architectures, such as WSRF [cOA04], introduce standards for connecting stateless Web services to stateful WS resources as shown in Fig. 1.1 and described above. Users can assemble software components (HOCs and others) from multiple services which store data across multiple invocations and can form distributed software entities which interact according to the recently specified Service Component Architecture (SCA [cOSO08]) which facilitates implementing important *quality of service* requirements: security, transactions and reliable messaging.

1.2 The Role of Middleware for the Grid

Middleware systems like Globus [bFo06] and Unicore [cUC08] are currently the most popular grid programming technologies. These systems form a layer between server operating systems and applications. A properly configured middleware automatically handles the exchange of data in the grid. However, programmers must provide the middleware configuration in the form of multiple files (declaring the data that is exchanged) written in middleware-specific *declarative* languages; these files must be arranged in a fixed order (declared in a file of its own) allowing the middleware to access them. The declarative code for middleware is not directly relevant for the application itself, it rather defines system properties, e.g., the data records stored by means of a WS resource (see Section 1.1), the assembly and encoding of messages and the server remote interfaces.

The use of a common middelware for all pieces of a distributed application brings the benefit of guaranteed interoperability at the price that programmers are required to handle the tedious middleware configuration, using declarative languages. Programmers using object oriented programming languages, where classes combine data and code, face the additional difficulty of object serialization: due to the widespread use of declarative languages which are based on XML (a data structuring and markup format [cW396]) in middleware technology, programmers must carefully consider how to transmit objects over the network, since they carry, besides data, executable code. When all these difficulties are addressed in the middleware configuration, the ratio between the amount of the problem-specific code of a program and the amount of its declarative configuration code quickly becomes so awkward that programmers tend to abstain from using grid middleware or the grid platform at all. For example, the introductory use cases in the Globus tutorial [cSC05] implement such trivial tasks as adding numbers: here, the actual computation is only a single line of Java code, while the configuration files span hundreds of lines of XML code.

The Higher-Order Components (HOCs) and the HOC-Service Architecture (HOC-SA) introduced in this book are an approach to simplify the use of modern grid technology by abstracting over the middleware. Moreover, they allow the programmer to benefit from tools providing more support for applications than only the component container which is the central program in most standard middleware systems. HOC programmers benefit from the Globus container which enables the HOCs to communicate over Internet boundaries, and, furthermore, from databases for sharing code and data, and from interpreters for combining code written in multiple programming languages. Advanced applications (discussed in Chapter 5) can also use an automatic task scheduler or an automatic loop-parallelizing compiler, without requiring from the programmer to directly deal with these tools.

Grid Computing

The potential processing power of the grid is greater than of any other high-performance platform since no supercomputer or traditional cluster can host as many processors as grid servers interconnected via the Internet. However, the top 500 list of the fastest supercomputers in the world is still led by homogeneous architectures. Not before 2006, when the Tokyo Institute of Technology launched the TSUBAME grid cluster [cTo06], grid technology was ever ranked among the top 10 positions.

Due to the grid's heterogeneity and its unsteady network connections, interprocessor communication in the grid is more difficult than in homogeneous architectures and, therefore, an increased number of processors does less likely lead to better application performance as, e.g., in a SMP server. The grid interconnects physically dispersed servers, and, thus, a grid platform is naturally a distributed-memory platform where any non-trivial application requires communication. The communicating servers in the grid have different hardware architectures, run different operating systems and the Internet connections they use for communication are of an unreliable nature, i.e., there are varying latencies and potential data delivery failures. Addressing the communication needs, therefore, poses one of the greatest challenges in grid application programming.

This chapter presents the communication technologies which are most popular for grid programming, starting from pure communication technologies (like

RMI & CORBA) to container software for hosting services and other components that benefit from the middleware in several issues: data persistence, logging, security, etc.

1.3 Communication Technologies for Distributed Computing

The traditional choice for handling the communication in high-performance computing (HPC) applications that target a distributed-memory platform are messaging libraries like MPI [cAl08] and PVM [bVa90]. Although approaches to porting these libraries to the grid exist (e.g., MPICH-G2 [bK+03]), applications using them are inherently low-level, since the grid-specific problems, e.g., converting data to a portable format, must be explicitly addressed in the application code.

In the following, some middleware systems are described, which can be seen as a complementary to the traditional HPC technologies. Using such middleware, the remote communication (across Internet boundaries) is no more the application programmers responsibility. While parallel processes on a local cluster or inside a LAN still communicate using, e.g., MPI, clients can trigger new computations and request results via the Internet. In data-intensive applications, the middleware can also be used to distribute data among multiple servers. Namely, RMI, CORBA and container software are discussed, focusing on the potential of such middleware for building an efficient component architecture.

1.3.1 Java Remote Method Invocation (RMI)

Remote Method Invocation [cSM07a] is an object-oriented communication technology for Java programs. Since RMI systems are built solely of Java programs on both the client and the server side, RMI provides portability for data and code: Any computer that runs a Java Virtual Machine (JVM) can be used as a server or client and Java's object serialization mechanism is implicitly used for encoding Java objects that are exchanged over the network, thus freeing the programmer from dealing with this low-level concern.

Figure 1.3 outlines the communication in an RMI system: so-called stubs (instances of Java classes, automatically generated from the user's description of the remote interfaces required for the application) handle all the network communication. All the RMI programmer has to do, is to specify the remote interfaces of the objects that are accessed over the network, i.e., the signatures comprising the names, parameters and return values of these objects' public methods. Remote interfaces are plain Java interfaces, with the only difference that programmers tag them, by deriving them from the base definition of a remote interface in the Java API and

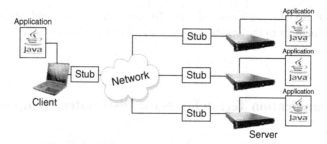

Figure 1.3 RMI Communication Model

declaring any method as a possible source of a remote exception (again, using a definition from the standard API). In Java versions below 1.4, there was a distinction between the client-side stub and the server-side skeleton and until Java version 1.5, programmers had to explicitly request the building of the automatically generated code from a specific RMI compiler (rmic). Recent Java VM versions support the ignoreStubClasses flag [cSM07c], allowing to create the stubs on demand at runtime, when set to true. The basic principle of all versions of RMI in Java is *location transparency*: there is no difference visible in the application code when either a local method or a remote method is called, since the encoding and decoding of data that is exchanged over the network (called *marshaling* and *unmarshaling*) is done by an automatically generated stub class.

The portability of Java code and the other RMI features including the standardized serialization mechanism for exchanging data and code, make RMI a candidate for programming grid applications. Some extensions to RMI that make it even more suitable for this purpose have been recently developed. Among the most notable are JINI [cSM07d], JXTA[cSM05c], Java Spaces [cSM07e] and Future-based RMI [bAG04]. While JINI, which simplifies the locating of other servers in the network, and Java Spaces, which provide a form of distributed-shared memory, and the Peer-to-Peer Technology JXTA, were not exclusively designed for the grid, future-based RMI is closely related to the component technology presented in this work: it introduces asynchronous method calls to RMI, speeding up RMI-based grid systems. An experimental grid system using future-based RMI has been developed, wherein recurring patterns of parallel behavior (skeletons) are provided as reusable implementations [bA$^+$02]. Similar research was conducted targeting parallel machines [aCo89], but the novel feature of the future-based RMI system is that it enables the use of multiple networked computers for one application.

While Java guarantees portability, the strong restriction to Java in RMI programs is also a shortcoming: legacy codes written in Fortran or C/C++ cannot be remotely accessed using RMI. Moreover, executing Java byte code on top of a Java VM and using its implicit serialization mechanism have some overhead as compared to native code which directly executes on the target hardware and allows programmers to use a more efficient messaging library without object serialization.

1.3.2 Common Object Request Broker Architecture (CORBA)

In Fig. 1.4, the basic setup of a CORBA-based distributed system is shown, which, in contrast to Java RMI, allows its users to interconnect programs written in almost any object-oriented language.

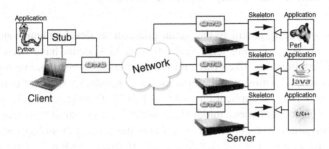

Figure 1.4 CORBA Communication Model

The Common Object Request Broker Architecture (CORBA [cCO04]) is a standard of the Object Management Group (OMG, the international computer industry consortium, also responsible for, e.g., the popular Unified Modeling Language UML). CORBA is used for communicating in object-oriented distributed systems that potentially combine programs written in different programming languages. The advantage of using multiple programming languages in a system composed of multiple programs is that each program can be written using the language that best fits its individual needs. Programs which are often executed on different architectures, e.g., can be written in a language which is compiled to an intermediate format (such as byte code) like Java and C# (both requiring a VM), while programs where sources are often changed can be written using scripting languages like Perl and Python (requiring an interpreter), whereas performance-critical programs can be written in a language like C/C++ (compiled directly into binary code for the target hardware).

The technical setup in CORBA is quite similar to RMI: local and remote objects are accessed in a transparent manner, while the marshaling and unmarshaling is not a part of the application code. While RMI programs are self-contained (clients and servers use stub objects, but RMI stubs are not programs of their own), all the computers in a CORBA system must run an Object Request Broker (ORB). In CORBA, there is also a distinction between client-side stubs and the skeletons handling the communication in the server-side programs.

The stubs and skeletons in a CORBA system are automatically generated (like RMI stubs), but instead of giving the interface description in Java or any other specific programming language, a language-neutral Interface Definition Language (IDL) is used, which is compiled by an IDL compiler. Besides marshaling and unmarshaling data, the ORB provides various services to the distributed system programmer, e.g., the management of distributed transactions. Thus, CORBA provides more features than RMI, but requires also more setup work for managing the ORB.

A notable approach to adapt CORBA for programming grid applications was developed in the Padico Project [bD$^+$03], where multiple MPI-based programs were connected into a distributed system using CORBA.

1.3.3 Containers for Components & Services

While grid computing is still in the initial phase of its development, valuable experiences with programming large-scale distributed systems have been made in e-commerce applications, like eBay and Amazon, which already serve millions of concurrent users. The most popular approaches to building a standard communication technology for grid applications, Globus [bH$^+$05] and Unicore [cUC08], have recently started to implement middleware which is quite similar to that used by business applications in e-commerce. Especially the concept of multiple application tiers, software components and containers [aV$^+$02] were obviously borrowed from e-commerce and adapted for grids.

Figure 1.5 shows the typical setup of a distributed system that uses an application server (called *container*) as its middleware.

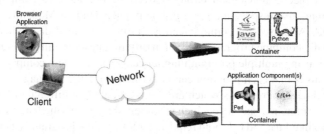

Figure 1.5 Application Server Communication Model

In the depicted example, the topology, i.e., the distribution of program parts onto physical hosts differs from the CORBA and RMI examples shown before. On the server side, only one program, the container which is a standardized part of the middleware (e.g., IBM's Websphere [cIB03a]) is running. The application-specific part in a server program using a container is not self-contained, but rather represented by one or many software components [aSz98] which realize application entities. E.g., a shopping cart component implements only the application-specific methods for purchasing some kinds of items. The server program which runs, e.g., multiple shopping cart components and connects them to an inventory database is embodied by the container. The client program does not necessarily need to be a particular program, but it is typically represented by a general-purpose Web browser like Firefox, especially in e-commerce applications, which all offer a client interface for ordering products by clicking on them.

The 3-Tier Application Model

The purpose of containers for application components is a simplification of the software development process. This simplification is achieved by freeing the developer from implementing their own communication (using, e.g., RMI or CORBA directly). Some containers also provide implementations for other low-level tasks that recur in many applications, e.g., database access. Typically containers provide programmers with a framework where components are classified into different types according to different application tiers. Many applications adhere to the popular three-part model-view-controller architecture (MVC [cRe79])) which roughly corresponds to setting up three application tiers as follows: a **presentation tier** (the view), a business or logical **application tier** (the controller) and a data **persistence tier** (the model).

Grid applications and Web applications both use the Internet as the underlying communication network. As will be shown in the following, there are more similarities between both application types. Both, e.g., have very low-level requirements, like the representation of some application data in structured data records and higher-level requirements, like running a certain parallel computation schema (e.g., divide-and-conquer). In the Web and in the grid, there are also application-level requirements, like rendering data as an image. These requirements are fulfilled by different parts of the system running the application, e.g., the middleware (embodied by the container), the employed software components or the user code. The following paragraphs describe each of the three tiers in the fundamental 3-tier paradigm commonly used in Web applications. A similar design is useful for structuring grid applications as well.

1. In the **presentation tier**, a recurring task is, e.g., rendering a part of the user interface in an HTML page and sending it to the user via HTTP. If this GUI part is, e.g., a text label and the employed container supports components of the type *Java Servlet* [cSM99], then programmers do not need to deal themselves with processing the HTTP GET-request (sent by the browser when the client accesses the page), but rather write a simple servlet component. This component is derived from the HTTPServlet base class and overrides its doGet-method, inheriting the base definition from the Java API. This class specifies that the doGet-method is provided with a response object connected to a PrintWriter, which allows programmers to send out HTML code that is rendered in the client's remote browser in the same simple manner as displaying some text on the local screen. Contrary to a plain HTTP server, e.g., Apache [cAo96], that only hosts static pages, servlet containers allow programmers to implement the presentation tier components dynamically, e.g., the page content is generated on demand, depending on the application state. The programmer's servlet code (e.g., the doGet-method) is automatically executed within a thread provided by the container. Thus, the server can concurrently process simultaneous requests, while Web programmers never have to deal with multithreading themselves. To enable the container to automatically serve requests by running some readily provided API code which embeds the programmers' custom code, programmers

must declare their components in configuration files and place these files in specific subdirectories of the container installation directory (a Java servlet is, e.g., described in a file called `web.xml` placed inside a directory called `webapps`). Thus, containers require users to write, beside component code, configuration files (typically, in XML-based formats). The process of making a component available in the network by providing the binary code plus the configuration file(s) inside the appropriate directory is called the component *deployment*. The grid component architecture HOC-SA, presented in Chapter 3 of this book, also provides a servlet-based presentation tier in the form of a portal.

2. In the **application tier** of a distributed system architecture, Enterprise JavaBeans [aMo06] are a well-established type of software components. Component containers (like EJB servers and the components deployed thereto) do not form an alternative to the middleware discussed above (RMI and CORBA), but they represent a complementary technology. Actually, many containers handle the communications using a combination of different middleware (especially in applications processing actions which are more complex than GET, POST and PUT requests in HTTP on the presentation tier). EJBs communicate via RMI using CORBA-compliant IIOP [cSM06a] as their wire-protocol, i.e., EJBs can be accessed from client programs written in other programming languages than Java. EJB programmers neither deal with RMI, nor with CORBA or IIOP, but simply declare a class, e.g., as a *Message-Driven Bean* (MDB [aMo06]) which is used for asynchronous messaging in EJB systems. By declaring that an MDB is responsible for processing certain messages, programmers can cause the server to execute this MDB's custom `onMessage`-method whenever a relevant message arrives at the container. This server-side communication is handled using the Java Message Service (JMS [cSM07b], an extension to RMI/IIOP for asynchronous messaging among EJBs). An interesting alternative to using EJB on the application tier is the Spring Framework [cSP08], where components communicate using either RMI, CORBA or SOAP. The grid component applications developed in this book also communicate via SOAP (Chapter 2) or a combination of SOAP with RMI (Chapter 4) on the application tier. Chapter 3 shows an example that makes use of SOAP and also MPI.

3. In the **persistence tier**, application components connect to databases using machine-oriented protocols. Typically, statements in the Structured Query Language (SQL [cSQ99]) are directly issued via TCP-sockets. In EJB systems, the persistence tier is represented by so-called *Entities* which have certain attributes automatically mapped onto relational databases by the container. For this purpose, programmers declare the relevant attributes as *persistent fields* and *relationship fields* [aMo06], reflecting the classical Entity/Relationship schema [bG$^+$76] of the underlying database.

Example Components: EJBs, CCM & .NET

On all application tiers, the component declarations are done using specific configuration files (XML-based ones for EJB). Besides the features necessarily required by the application, some containers offer also enhanced features, e.g., an automatic load balancing: programmers can define a threshold for the maximum number of component instances per server. When this threshold is exceeded, the next instance will be created by a container hosted on another machine in the network. While the features offered by container software differ from product to product, all containers provide the functionality of running custom components within a readily provided framework that frees the programmers from implementing the same application functionalities again and again.

Besides the popular EJB technology, there are other component technologies using container software, e.g., .NET [aLo01] and CCM [cCO07], allowing programmers to use other languages than Java for implementing their components. Examples of popular containers are, e.g., WebSphere [cIB03a] and WebLogic [cBl06], which are often called application server software by their vendors.

Accessing Components via Web Services

The recent trend in grid middleware (e.g., Globus and Unicore) is to adopt the design concepts from e-commerce applications and run HPC components inside containers of a specific type: *Web service* containers. A detailed discussion of using Web services in grid computing follows in Chapter 3. The most crucial factors which motivate the use of Web services for communicating among grid components are the three most characteristic features of this communication technology:
Web services

1. ...provide recurrently used application code in the form of *operations*, which allow the user to run an application activity on a remote machine;
2. ...are deployed into containers together with a configuration (similar to the components presented above) which enables the container to handle the network communication when an operation is requested remotely. This configuration comprises two files per service:
   - the **deployment descriptor** file, **WSDD**, which defines (in an XML-based format) what code file implements what operation,
   - the **interface definition** file, **WSDL**, which is an XML-based IDL file, defining all the operations' input and output;
3. ...use standardized XML-based messages in the **SOAP** format for encoding the requests for operations and their responses.

The use of XML for all communication-related declarations (and for the request/ response messages themselves) allows to use Web services for interconnecting programs which are written in various programming languages and run on various architectures. This feature is often referred to as *interoperability*.

Contrary to the multiple procedures or methods assembled in a class, module or other composite code which is technically connected (e.g., by sharing a common variable space), the operations assembled into a single Web service form a purely conceptual union of technically separate, typically *self-contained* code. Therefore, no continuous connection between the client and a specific service instance is required, but it makes no difference on the client side, if subsequent operations are requested from the same physical service or from different instances of the service which may be deployed on multiple servers (this is the *loose coupling* principle [aKa03]).

Most modern container products that allow to host components, such as the EJBs presented above, allow the users to expose these components via Web services automatically. Thus, programmers can create and deploy Web services, which offer the required operations for accessing remote components (e.g., an EJB) by clicking a button in a configuration tool (e.g., an IDE like Eclipse [cEc03]) and never need to deal with WSDL and WSDD files.

Web Services and the Grid

As already mentioned, modern grid middleware (e.g., Globus and Unicore) also comprises Web service containers. Due to their interoperability and loose coupling features, Web services may seem to be the ideal technology for interconnecting grid components. However, the components required for programming grid applications are fairly different from those used in e-commerce. The future-based RMI system discussed above [bA+02], e.g., may be considered a grid component system. In contrast to e-business components representing entities like shopping carts or items for sale, which have little dependencies among each other and can be almost arbitrarily managed by multiple threads (on possibly multiple networked servers) by creating new instances, each single grid component typically spans multiple servers. In HPC applications, it also must be taken into account that marshaling and unmarshaling SOAP messages is quite time consuming, and, thus, Web service communications should be reduced as much as possible.

Even the most simple parallel processing patterns which are sometimes referred to as *embarrassingly parallel*, since they can be divided into independent tasks, are not trivial to implement as grid components. An example is the popular *compute farm* [aCo89] pattern where one *master* process divides the input data of an application into multiple tasks and distributes them among multiple *worker* processes. An efficient implementation of a farm component for the grid requires that multiple communication technologies (e.g., a Web service container and RMI servers) are combined together. The reason for these complex demands is that there are two different kinds of communication in a grid-enabled compute farm:

1. **client/master**: this communication is only required twice for each application:

   a. when the application is launched and the input is uploaded,
   b. after the results have become available and are downloaded.

   Thus, it does not strongly matter whether a very efficient communication technology is used, but rather features like interoperability and loose coupling are important. Using a Web service for this purpose makes the farm component available to a broad range of clients and allows clients to easily interchange different farm implementations in different applications.

2. **master/worker**: this communication is required for every input and output item. To ensure that the speedup gained by running the application in parallel is not undone by the communication overhead, an efficient communication technology must be used. Moreover, once a more complex farm-like component, where tasks are not only described by data but also by code (i.e., not all workers do the same), purely XML-based communication is no more sufficient, but rather a technology that supports code transfer (e.g., RMI) is necessary.

Figure 1.6 shows a possible setup for a grid application that uses a distributed compute farm component, made accessible via a container software.

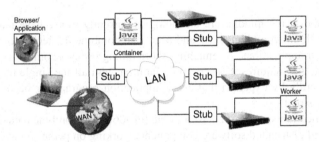

Figure 1.6 Communication in the grid for the Compute Farm Pattern

In the depicted example, the container hosts only one Java component (the number of components hosted by a container can be set up by the user according to the application requirements). For a farm, the most effective setup is probably to host only the master inside the container (as a central access point, allowing a broad variety of clients to connect), while the interconnection among master and workers (which is typically established before an application runs and contains a fixed number of unvarying machines) can use a more tightly coupled (i.e., better performance, but less flexibility) communication technology (RMI, in this example). This topology makes sense also for other grid components, different from the compute farm. Indeed, most of the commonly known design patterns in HPC programming [aRG03] have internally a fine-grained structure, which is best implemented on a cluster or in a LAN. In a LAN environment, libraries like MPI also do not cause conflicts between communicating servers, since the same version of the library can easily be installed on each server in such a *subnet* of the grid. The container will probably be represented by the Globus container [bH+05] in this setup. The popular Globus Resource Allocation Manager (GRAM [bC+98]) actually suggests a similar approach: a Web service, called WS-GRAM, is used to submit programs which communicate using, e.g., MPI. But in the WS-GRAM approach to grid programming, the submitted programs do not benefit from reusable components provided on the server side. Instead, the submitted programs are responsible for handling all application requirements.

Whether an application uses GRAM or not, the Globus container is an adequate middleware for many different grid applications. Besides Web service hosting, the Globus container offers features like asynchronous communication (i.e., the client will not be *busy waiting* while computations in the grid are running) and the maintenance of an application state (using WS resources [cOA04]). This state is represented by data which outlasts a single Web service operation, i.e., in Globus, the Web service operations are not necessarily all self-contained. This feature is necessary for a parallel implementation of most Web service-based applications with dependences, i.e., some operations can affect the behavior of other operations.

1.4 Shortcomings in State-Of-The-Art Grid Middleware

As shown above, the middleware technologies currently in use have many useful features which are complementary and users can combine the technologies, as required in their applications. A remaining shortcoming is that a combination of multiple technologies requires its users to be familiar with all the single technologies and it can hardly by expected that every application programmer becomes a grid middleware expert.

Web application servers, e.g., Web Logic [cBl06], also combine databases, Web interfaces and distributed software components working on persistent and transient data. However, grid middleware systems have even a steeper learning curve, since most of the tools which simplify Web application development (e.g., the numerous plugins [cEc03] for the popular Eclipse IDE) cannot be used for grid application programming. Unfortunately, Web application development tools have little to offer for applications in grid architectures (like the one depicted in Fig. 1.6) where large parts of the system run outside the container which only serves as an access point. Thus, despite of the rapid progress in grid middleware development, the lack of reusable components and tools for composing them into applications remains a severe shortcoming of this technology.

1.4.1 Responsibilities of the Middleware User

When a grid application, offering all the features listed above (remote access to its components via a Web services, locally enhanced communication, etc.) should be developed, the programmer who directly uses the middleware has to do much more than for implementing a traditional HPC application:

1. the application logic must be implemented, which requires the programmer to be an expert in the application domain (e.g., pattern recognition);
2. the remote component for processing the application in parallel must be implemented, which requires the programmer to be an expert in parallel processing strategies and the employed communication technology (e.g., RMI or MPI);
3. the Web service used to access the component in the grid must be deployed, which requires the programmer to be an expert in middleware technology and the XML-based formats used for the container configuration.

Even the latest version of the popular Globus container requires the programmer to do the deployment of any grid application manually.

1.4.2 Requirements for the Software Components

Application programmers should reuse existing components and implement only the application-specific parts of their programs. Like a communication library for a certain platform type (e.g., MPI), an adequate component model for the grid must be versatile to address the different needs of possible applications. Additionally, software components for the grid must comply with the common middleware standards, allowing remote clients to access them in a standardized manner.

This chapter has shown that different middleware systems serve different purposes, e.g., a Web service hosted by the Globus container can neither guarantee good performance nor scalability. The main advantage of this technology is that it interoperates flawless with most other software in the grid, due to the use of portable data formats. The efficiency of the data processing depends on the implementation of the service operations. Thus, it remains either the programmer's duty to handle all the data processing, or software components are required that can be easily adapted to different applications, while the required middleware support for accessing them remotely is already included. The following chapter systematically introduces Higher-Order Components (HOCs) which satisfy these requirements.

Chapter 2
HOCs: Software Components for Grid Programming

This chapter introduces the main contribution of this book – *HOCs – Higher-Order Components* – that provide the grid application programmer with reusable and composable implementations of recurring parallel processing patterns. HOCs [bGD05] can be viewed formally as higher-order functions: a generic implementation of a HOC on a remote machine can be customized with application-specific code parameters which are supplied by the user and shipped via the network.

The motivation to use software components for grid programming is driven by the special features of this platform type: in the grid heterogeneous, high-performance resources are connected via the Internet. The main appeal of grid computing is the easy, seamless access to both data and computational resources that are dispersed all over the world, regardless of their location. However, initial experience with the grid has shown that the process of application programming for such systems poses several new challenges.

While the grid basically originates from parallel systems, grid programming adds a new level of complexity to the traditionally challenging area of parallel programming. The main reason for this is that the gap between the system and the application programmer has grown with the transition from parallel to distributed systems and further to the grid. Low-level parallel programming approaches such as MPI require the programmer to explicitly describe communication between processes by means of a suitable message passing library. Distributed systems and the grid require in addition that software interfaces accessible remotely via network connections are declared explicitly. The current technologies for doing this (CORBA IDL, RMI, DCOM/COM, WSDL, etc.) are numerous, but unfortunately either very complex or non-flexible (as discussed in Chapter 1). This book shows that components can free the programmer from having to explicitly deal with all the underlying technologies of the grid. To motivate Higher-Order Components (HOCs), this chapter points out the benefits of using (besides data) code as a type of parameter which Web clients can supply to a component.

The structure of the chapter is as follows:

Section 2.1 motivates the use of components for grid programming and introduces HOCs. This section systematically subdivides the code of a grid

J. Dünnweber, S. Gorlatch, *Higher-Order Components for Grid Programming*,
DOI 10.1007/978-3-642-00841-2_2, © Springer-Verlag Berlin Heidelberg 2009

application into an application-dependent part and an application-independent part. The application-dependent part is implemented anew for each program by the *application programmers*. This part is where the application logic is defined. The application-independent part is encapsulated by grid *system programmers* into a component that is reused in many different applications.

Section 2.2 illustrates the benefits of using software components by a comparison of programming one sample application twice: once without components and once using HOCs. The well-known "Farm of Workers" pattern is taken as a motivating example. This section presents an implementation of the Farm-HOC, such that the component is accessible via a Web service hosted on top of the Globus middleware. As a first case study, the Farm-HOC is used for computing fractal images, and the performance of this application is experimentally evaluated using a testbed consisting of distributed high-performance servers. This section focuses rather on the configuration code required by modern grid middleware than on the application-dependent code. The detailed discussion about middleware configuration makes it apparent that the distribution of programmer roles on the grid into application programmers and system experts is advantageous, since all configuration code can be made application-independent.

Section 2.3 describes the Java-based application programming interfaces (APIs) for the two different grid programmer roles: While the client-side API offers simple adapter classes for invoking HOCs via Web services (as already shown for an example in Section 2.1), this section explains in detail the server-side Service API. It is shown how new HOCs can be derived by extending generic parent classes and inheriting the base definitions from the Service API.

Section 2.4 addresses the important problem of HOC adaptation: while the relatively simple Farm-HOC is suitable for running any dependence-free application on the grid, it does not cover applications with data and/or control dependences. We demonstrate how the standard behavior of a HOC can be modified, such that, e.g., the Farm-HOC takes dependences into account. Using an example application with so-called *wavefront* data dependences, it is shown how adaptations help to save a lot of programming work. Instead of developing a new HOC for every kind of dependence (of which a multitude exist), an existing HOC can be adapted via its code parameters.

Related Work section is focussing especially on the relation between component adaptation and aspect-oriented programming.

2.1 Higher-Order Components (HOCs)

Higher-Order Components (HOCs) address the important and difficult problem of simplifying the programming of grid applications. The foremost goal is to provide a high-level programming model which shields the application programmer from the low-level details of the heterogeneous and highly dynamical grid environment, thus allowing him to concentrate on algorithmic and performance issues.

The HOC concept originates from the following two veins of previous research: (1) component-based software development [aSz98], with the major goals of reuse and compositionality, and (2) skeleton-based programming [aCo89], aiming at identifying and abstracting typical patterns of parallel computing. HOCs are a novel kind of components which can be parameterized with application-specific code. The use and implementation of HOCs is demonstrated on examples in the contemporary grid context of OGSA/WSRF [cOA04] and the Globus Toolkit [bFo06] (OGSA is the implementation of the resource management and communication standards WSRF [cOA04] and WS-N [cOA06] in Globus).

2.1.1 Motivation for HOCs

Components (as described, e.g., in Szyperski's book [aSz98]), capture common programming patterns as independent, composable program units and present a high-level API to the application programmer, hiding hardware-specific details. The benefit of a clear software development organization, due to the division of platform-specific and application-specific programming into two separate work processes, is typically referred to as the *separation of concerns* in component-based software development. Another important advantage of using components is code reuse: different applications requiring common functionality can share a common component implementing that functionality. The same applies to the middleware setup which consists of writing various configuration files and placing them properly on the servers. Once a component is properly set up, different applications can benefit from this setup, without requiring any changes to the configuration files. The purpose of setting up components by configuring the middleware, is that system-level issues are delegated to the middleware and not included in the application code. These issues include the network communication and, in many middleware systems, also system-level features which are useful, but not obligatory for running the application (data persistence, security, etc.).

Basic requirements on a piece of software in order to be a component have been gathered in the CoreGRID definition of a component [cCN05]: "A component can be viewed as a black box, for which the specification of its interface to the external components is sufficient. More precisely, each component defines a set of ports (also called interfaces) that define the type of the component. Ports can be connected (bound) together provided they are compatible. The compatibility relation must be defined at the level of the component model and must be verifiable both at composition time but also at runtime." Technically, compatibility among components is provided, when the output produced by one component is a valid input for another component, and interface declarations exist, such that this property can be automatically tested. HOCs fulfill this requirement and, thus, can be viewed as one possible implementation of the CoreGRID component model (GCM).

Application programmers get most benefits from HOCs when the separation of concerns corresponds to a separation of programmer roles: while application

programmers only write code parameters, the HOC's implementation and configuration can be developed by a grid system programmer. In this model, application
programmers only deal with application logic, and grid system programmers are
concerned with optimally exploiting the processing power of the parallel servers.

2.1.2 Grid Programming Using HOCs

To illustrate the idea of grid application programming using HOCs, three example HOCs are depicted in Fig. 2.1. Each HOC has a generic structure which can
be reused in many applications. This feature was adopted from skeleton-based programming [aCo89]. The Farm-HOC (Fig. 2.1, left), e.g., implements the popular
farm pattern which is used for the first example application in this chapter. The
Farm-HOC is an example for a data-parallel component [aG+02], as all parallelism
in the farm is implemented by applying the same code to different data at the same
time. This book deals with data-parallel and also with task-parallel HOCs. An example for a task-parallel HOC is the Pipeline-HOC (Fig. 2.1, right) which executes
a sequence of functions, each corresponding to an application stage. Each stage performs a (potentially) different task in parallel with the other stages for multiple input
instances; therefore, the term task-parallel. Similarly, the Wavefront-HOC (bottom
of Fig. 2.1) also has multiple stages, but there is a varying degree of parallelism on
each stage.

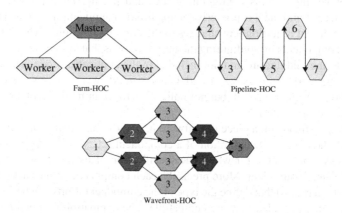

Figure 2.1 Processing Structures Implemented by Example HOCs

The Farm-HOC has the simplest parallel computation structure of all HOCs. The
function computed by the Farm-HOC workers (and the manner in which the input
is partitioned by the master) depends on the application. Therefore, this is specified by the application programmer who provides for this purpose executable code
as parameters to the HOC. These code parameters are the reason, why HOCs are

called higher-order components, like higher-order functions that take functions as arguments.

The benefit of using code as parameters for components is a high level of abstraction: programmers only need to care about *what* to compute in parallel and express it via the code parameters. *How* the parallelism is actually achieved, is expressed inside the component code that is readily provided to the programmer. Once a code parameter has been uploaded to the so-called Code Service (a Web accessible database, explained in detail in Chapter 3), it can be used in many applications. Thus, HOCs combine the advantages of skeleton-based programming [aCo89] with the advantages of pattern-based programming (as suggested by Gamma et al. [aG+95]). While patterns provide to the application programmer only recipes for solving recurring problems efficiently, skeletons provide fully implemented solutions. Typically, a skeleton is implemented differently for each particular kind of parallel architecture. HOCs, in contrast, are not coupled to a certain execution hardware but rather to a middleware which can run on many different hardware platforms and can be accessed remotely from any Internet client. In the implementation of all currently available HOCs (see Section 3.2.3), the Globus toolkit [bFo06] is used as the middleware, and Web services [cW302] are used for communication.

Each HOC consists of two parts: (a) a set of interfaces specifying the signatures of the HOC's parameters, and (b) a server-sided component implementation. Since code parameters for a HOC carry method implementations, the signature of each parameter (method's name, types of the arguments and the result) must be provided by the HOC developer to enable the method implementation for exchange among networked computers.

```
1:  interface FarmHOC<E> {
2:    setMaster(int masterID); setWorker(int workerID);
3:    E[] compute(E[] input);   }
4:  interface Master { E[][] split(E[] input, int numWorkers); }
5:  interface Worker { E[] compute(E[] input); }
```

Figure 2.2 Interfaces for the Farm-HOC

Figure 2.2 shows the example interfaces for the Farm-HOC in Java notation. The type parameter <E> denotes that the input can be any data which can be split into discrete elements which are distributed among numWorkers computers. There is one interface for accessing the component remotely (lines 1–3) and two code parameter interfaces:

(1) the Master (line 4) that splits the input data in an application-specific manner for distribution among the workers, and
(2) the Worker (line 5) that processes a unit of data in an application-specific manner.

Lines 2–3 in the above code show, how the Farm-HOC is remotely accessed: the code parameters used in an application are selected via the appropriate identifiers (masterID and workerID in line 2) and compute (line 3) starts the remote processing of data (of the variable type E). In a concrete implementation, the Farm-HOC can be realized in a hardware-specific way, for handling the parallel computations as efficient as the runtime platform allows. The middleware is responsible for making any implementation accessible in a platform-independent way, via the uniform Farm-HOC interface. Thus, many different implementations of the Farm-HOC can be provided by the grid system programmers, each tuned to a particular hosting architecture (e.g., a multithreaded version for shared-memory workstations and a version based on Java RMI or a messaging library, such as MPI [cAl08] or PVM [bVa90], for distributed-memory clusters), allowing the application programmers to select among them (depending on available execution hosts and application requirements, such as memory needs, connection performance, etc.).

The client-side implementations of the Master and Worker parameters (according to the interfaces in lines 4–5, above) must be written in an application-specific, hardware-independent manner.

The Java code in Fig. 2.2 is only for illustration purposes and does not necessarily look exactly so in an implementation. The interfaces of a HOC can be described using a different notation than Java. Actually, most implementations in this book use WSDL [cW302], since Web services are used for remotely accessing the HOCs. The interfaces shown in Fig. 2.2 are simplified: more detailed descriptions follow in Section 2.3 where the corresponding WSDL code is shown.

To develop an application using HOCs, the application programmer first identifies the HOCs suitable for the application and expresses the application using the selected HOCs and their parameters. When the application is executed, grid servers executing the HOCs (e.g., Worker code) are selected at runtime, either according to a default setting, or via an automatic scheduler as shown in Chapter 5, or by the user as shown in Section 2.3.

2.1.3 Introducing Code Mobility to the Middleware

In the following sections, an example implementation of the Farm-HOC is presented that uses the Globus Open Grid Service Architecture (OGSA [bFo06]) as the middleware for interconnecting the client with the master and the workers. Since the central technology in the OGSA is a Web service container, all the network communication in this implementation is handled using Web services.

Because HOCs are implemented on grid hosts while the application-specific, customizing code parameters reside on clients, HOCs require facilities for code mobility, i.e., shipping code from clients to servers and executing it there. Code mobility mechanisms are currently available in Java/RMI-based distributed computing technologies like Jini or JXTA. However, the communication protocols used by RMI are often unable to pass through firewalls and Internet-proxies in a grid environment. Additionally, using RMI for communication would reduce the programming language choice to Java.

In contrast, OGSA [bFo06] is suitable for Internet-wide applications, allowing communication across firewalls and proxies. Standardized XML-based formats are used for the remote procedure calls, the remotely accessible data records and the descriptions of the appropriate interfaces for the remote communication, namely SOAP [cW302] and WSDL [cW306]. Therefore, it is possible to combine code developed in different programming languages using Web services.

Due to the universality of OGSA systems, no language-specific format can be used for declaring the type of a piece of code that is used as a parameter (e.g., the Java interfaces in Fig. 2.2 or a Java class type). Such a declaration constraint is not only present in Globus OGSA, but in any programming language-neutral middleware (e.g., Unicore [cUC08]). Since WSRF/OGSA in Globus do not define any standard implementation for code mobility, we extend the Globus middleware and introduce a novel code mobility mechanism.

In comparison to RMI, the SOAP communication mechanism used by Web services is more restrictive: it is not possible to pass an object of a type that cannot be described using the XML-schema language [cW396]. Therefore, SOAP parameters may consist of primitives or some kind of data record (possibly declared in a class-like manner, e.g., pure Java Beans [cSM97] which are classes with trivial methods that can only set or get attribute values), but once a non-trivial implementation (i.e., an algorithm) should be transferred, this mechanism is not sufficient.

To introduce the code mobility mechanism for HOCs, Fig. 2.3 shows the transfer schema for the Farm-HOC on a high level (Chapter 3 explains in detail, how code mobility via Web services for any HOC works). At this point, only the basic issues are briefly addressed:

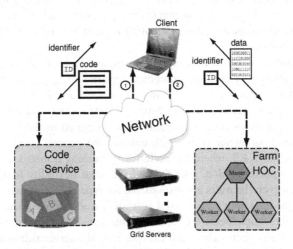

Figure 2.3 Code Transfer Schema for the Farm-HOC

① the application programmer stores the implementation of code parameters for HOCs to the Code Service which returns one unique identifier per parameter, allowing the programmer to refer to the remotely stored code later on;

② the client starts the remote computations by sending to the HOC a request which contains the input data and the identifiers of the code parameters to be used for the data processing.

Both, ① and ② are Web service calls. The identifier (ID, which is the output of call① and the input of call ② in Fig. 2.3) is critical for enabling code mobility: HOCs and clients use such identifiers, which are implemented using string literals or numeric primitives, to uniquely refer to code parameters. This referencing mechanism enables the transfer of the application-specific parts in an algorithm to a Web service as if they were plain data (i.e., the code is declared as a data array in the Web service interface). By referencing to a code parameter via an identifier, users determine what type of code parameter (e.g., a farm worker) is referenced and, thereby, how to convert this data (the character array) back into executable code on the server side.

2.1.4 Polymorphism and Type Checking for Code Parameters

If a code parameter is written in an object-oriented language, it can be converted on the server side using a type cast operation that directly assigns to the parameters the appropriate class or interface types (e.g., the Master and Worker interfaces in lines 4–5 of Fig. 2.2 for the Farm-HOC). In C++, pointers to the code (represented via data arrays) are converted into pointers to functions, also via type cast operations. While C++ allows to apply a type cast to an array pointer for achieving a function pointer, Java requires reflection for creating an instance of the class described by the code, but these details of converting code are transparent to the Code Service users. A non-object-oriented code parameter rather requires a certain command string to be assembled for executing it (as shown in the C/MPI Gateway example in Section 3.4.1).

Both, Java and C++, allow for polymorphism, and type casting of code parameters which enables very general parameter declarations (e.g., the Master and Worker interface types from Fig. 2.2 in case of the Farm-HOC). Users are free to use more specific interface or class types, as long as they provide compatibility with the general types (i.e., implement all the required methods). An example of a specific code parameter type which extends the general Worker definition is shown in Section 2.4.2, where the code for processing a matrix is defined. By granting polymorphism to HOC users, they have great flexibility in defining code parameters. However, the users are responsible for type checking: when a code parameter of an incompatible type is used (e.g., if a worker code parameter for the Farm-HOC neither implements the Worker interface nor an interface that is derived from Worker), then a class cast exception on the server side occurs. In case of a non-object oriented code parameter, users also need to perform a careful type checking, since an execution failure on the server can result from using code parameters which do not define syntactically correct command strings.

The parallel HOC implementation has gaps which are filled in by the application-specific code parameters. These gaps are represented by abstract descriptions of the code parameters in the form of interfaces, abstract class types or command strings with placeholders. The identifiers, which the clients use for referring to code parameters make clear where each code parameter belongs to on the server side and what operation (a class/type cast or command string concatenation) is required to make it executable there. Once all parameters have been transferred to a HOC via the Code Service and properly converted, the code is linked to the server-sided HOC implementation and the execution of the HOC can be initiated.

The proposed code mobility mechanism assumes that the code for a HOC parameter is runnable on the HOC host (at least, after a well-defined conversion). While our first example HOCs and applications in this book only use portable Java byte code, Section 3.1.1 addresses mobility of multiple different code formats (including non-portable ones).

Services and Factories for the Farm-HOC

There are different means for implementing the communication between the master and the workers in the Farm-HOC. In Chapter 1, the advantages and disadvantages of different network protocols and middleware systems were discussed.

As a proof-of-concept implementation, a version of the Farm-HOC is shown in this section that uses a Web service for connecting the client to the master server and also uses Web services for the communication between the master and the workers. Due to the relatively costly XML encoding process required by Web services (even primitive data must be wrapped into a tree-structured document [cW302]), this version is probably not the most efficient implementation option of the Farm-HOC. Which implementation version offers the best performance varies from application to application and also depends on the data being processed. In Section 1.3.3, we discussed using different middleware for the different kinds of communication in a farm. However, the version that uses only Web services has reduced complexity as compared to combining multiple communication technologies and is, therefore, used as the introductory example.

The presented Farm-HOC implementation consists of two different types of Web services:

(1) a Farm Service, of which a single instance distributes the computations for the Farm-HOC; it corresponds to the master in the compute farm model;
(2) a Worker Service, which is instantiated several times to perform a single computation, correspondingly to the worker's role in the model.

Figure 2.4 shows the computation and communication steps in the Farm-HOC processing: the client obtains a Farm HOC instance by requesting the required resources (explained below) from the farm resource home ①. To start calculations, the client communicates with the Farm Service ②. The Farm Service obtains the resources required by the Worker Services from the corresponding worker resource homes ③, creates and distributes the sub-tasks of the calculation ④ and reassembles the result which finally is returned to the client ⑤. This way, the Farm Service

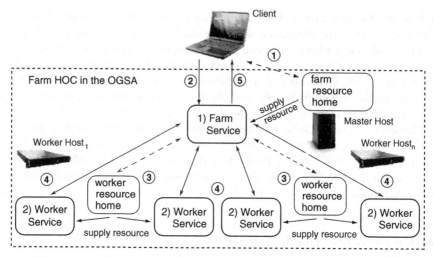

Figure 2.4 Farm-HOC Processing in OGSA

hides all the distributed interactions in the grid and provides a lean interface (an implementation of the façade design pattern [aG⁺95]) for the complete Farm-HOC. The machines in Fig. 2.4 show one possible network topology (i.e., a distribution of Web services to physical machines) which is typical for many applications: the client does not run any service, the Farm Service (including the farm resource home) is hosted on a machine called Master Host and the Worker Services (including the worker resource homes) are hosted on multiple machines called Worker Hosts.

As can be seen from the figure, for each Web service (the Farm Service and all Worker Services) there is a resource home which supplies the service with re-sources. The purpose of these resources (also called WS-resources) and the resource homes is the handling of a distributed application state. According to the WSRF standard [bFo06] for grid programming, *resource homes* implement the factory de-sign pattern [aG⁺95] allowing the user to request the creation of new resources over the network. Each resource is a data record stored on the server side. The relation between servers and resources is one-to-many, i.e., each resource is placed on ex-actly one server, but each server can potentially host many resources. For every resource that a resource home creates, one *endpoint* [cAp06] (omitted in the figure, since there is one endpoint transferred in every interaction) is returned to the client, allowing the client to refer to the resource later on. Using the remote resources, clients maintain the application state by specifying which resources are affected by which service operation. This specification is done via the endpoints, i.e., instead of exchanging all data that is processed by the services as input and output of each operation, the data remains on the server side as long as the application runs. Clients refer to the data (and eventually retrieve it) using the endpoints. For different kinds of resources, there are different resource homes (the example Fig. 2.4 uses one farm

resource home and multiple worker resource homes) which are specified during the
service deployment in the middleware configuration.

The presented implementation of the Farm-HOC provides a solution to the following problems:

(a) parts of the computational state (e.g., intermediate results) should be stored in
    the local memory of the machines where the results are actually computed (for
    reducing communication costs), and
(b) the client and all servers can access the data on other servers using endpoints,
    which decouples the machines: not all the concrete implementation code and
    data needs to be present on all the involved machines.

The Farm-HOC benefits from these two features of the presented implementation
for enabling asynchronous communication and, to a certain extent, adaptability as
follows: neither the client nor the master ever need to wait idle for computations, but
any other activity which does not require the full result of all worker operations can
be executed asynchronously with the worker activities. Moreover, the implementation code on some hosting platforms can vary from the others, e.g., only a single
worker implementation can be optimized for a server with a big cache size without
affecting the master since all workers share the same Worker Service interface.

Parallelism in the Farm-HOC is achieved by starting (at least) one new thread
(taken from a thread pool) for each worker, which sends back a notification to the
Farm Service upon completion of the calculation.

Despite the simplicity of the Farm-HOC, its implementation and configuration
on top of the Globus middleware still requires several laborious steps which are
difficult for an application programmer: defining service implementations, remote
interfaces, and setup (e.g., where to find what file), all using Globus-specific formats
(Section 2.2.1 explains each of these steps in detail for a concrete case study). It is
an advantage of HOCs that these steps do not have to be done repeatedly by application programmers, but rather accomplished once for each HOC by a grid system
expert.

```
1: farmHOC = farmHOCFactory.createHOC();
2: farmHOC.setMaster("masterRef");
3: farmHOC.setWorker("workerRef");
4: farmHOC.configureGrid( "masterHost",
5:                        "workerHost1",... , "workerHostN" );
6: farmHOC.compute(input);
```

Figure 2.5 Invocation of the Farm-HOC

Figure 2.5 shows how the application programmer uses the Farm-HOC in the
client code. The application starts by requesting a new instance of the Farm-HOC
from the farmHOCFactory (line 1 in Fig. 2.5). Lines 2–3 pass references to the
application-specific code parameters (as specified by the interfaces in Fig. 2.2) in the
Code Service to the HOC-instance. In lines 4–5, the execution hosts for both master

and workers are explicitly selected (which is an optional step, see Chapter 5). The HOC is then invoked to process input data on the grid (line 6).

2.1.5 First Application Case Study: Julia Sets

As a first example application of the Farm-HOC, so-called *Julia Sets*, which are a special kind of fractal image, are calculated [aPR96].

To compute the desired images described by Julia Sets in parallel, the rectangular computation region (initially, a blank matrix) is split into several tiles, mapped to a certain region of the complex number plane and then, a given function is applied to each number. The Julia Set diagrams used in the mathematics of chaotic systems are obtained using a polynomial function, e.g., $z \rightarrow z^2 + c$, whereby any c results in a different set. By iterating the chosen function, a sequence of numbers is produced that either diverges or converges; input numbers producing converging sequences are assigned to the Julia Set and the degree of growth in the associated sequence determines the color assigned to a particular point in a tile. Accordingly, there is an infinite number of Julia Sets, but the coefficient c must be chosen within the so-called stability range to derive a non-empty set.

Since all numbers in the examined section of the complex number plane can be processed independently, this procedure can be applied to all tiles in parallel, which directly corresponds to the farm pattern of parallelism. A specific feature of the farm in the Julia Set application is that the computation is a dynamic process which requires different amounts of time for different numbers, so, a notification mechanism is required to decide when the computation for a single tile is finished.

Using the Farm-HOC

The calculation process can be applied to each point independently, which facilitates a straightforward parallelization by dividing the plane into p rectangular tiles and distributing the computations among p processors. To implement this schema using the Farm-HOC, the application developer needs to provide a master and a worker implementation (the latter is shown in Fig. 2.6).

```
1: public class FractalWorker implements Worker  {
2:  public Object compute(Object tile) {
3:    for (int y = 0;  y < tileHeight;  ++y) {
4:      for (int x = 0;  x < tileWidth;  ++x) {
5:        ... // compute julia value for (x, y)
6: }} ...}
```

Figure 2.6 Example Worker Parameter for the Farm-HOC

The code shown in Fig. 2.6, which is uploaded and stored in the Code Service, is referenced to by the `WorkerID` argument in line 2 of Fig. 2.5.

The Farm-HOC design using worker factories allows for a kind of automatic load balancing: although a single thread is enough to execute the above `FractalWorker` loops, there is a version (used in the experiments below) which runs multiple threads, so it can process parts of a tile in parallel itself. Now, if the tasks are unequally balanced on a compute node, a worker thread that has been suspended early (which happens in regions where the computed sequences quickly diverge) will immediately free resources for another thread which is processing a more time-intensive task. Moreover, even for an equally balanced distribution of tasks, the multi-threaded worker version better exploits the available resources when it is deployed to a multiprocessor server, as each thread is mapped to a processor of its own.

2.2 HOCs and Grid Middleware

To demonstrate how much the role distribution between application and system programmers (suggested in Section 2.1.1) reduces the work that grid application programmers must do, the case study of fractal image computation from the previous section is revisited twice in this section: firstly, the application is implemented from scratch without the use of components and, then the implementation using the Farm-HOC is reviewed and compared to it. Particular focus is given to the middleware configuration which can be pre-packaged with the HOC, thus freeing the application programmer from dealing with any configuration files at all.

2.2.1 An Analysis of the Requirements of the Grid Platform without Components

The Julia Set application from the previous section, where the focus was on the Java implementation, is now used as the example of writing a middleware configuration.

Without the Farm-HOC from Section 2.1.4 available, the straightforward approach to implementing the considered case study on the grid is to develop specific Web services for computing Julia Sets (i.e., to "hardwire" the application). The most obvious difference between the generic implementation in Section 2.1.4 and a hardwired version, is that the latter requires the application programmer to explicitly deal with the implementation and the middleware configuration. To illustrate this difference, in the following, a hardwired implementation is presented with a focus on the required middleware setup.

For performing the Julia Set computation, the hardwired implementation uses two Web services hosted by multiple servers running the Globus middleware, as shown in Fig. 2.7:

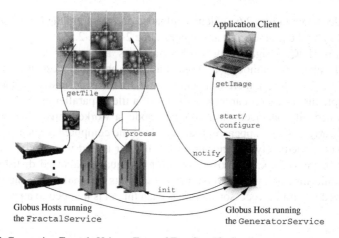

Figure 2.7 Computing Fractals Using a Farm of Two Specific Services

(1) the GeneratorService, that provides the public operations configure, start, notify and getImage, and
(2) the FractalService, that provides the public operations init, process and getTile.

These services work together as follows:
to set up a collection of servers calculating a fractal image, a GeneratorService instance is requested from one server, and the client provides a list of servers that will compute the single tiles using the configure operation. The configure operation requests a FractalService instance on each specified server and calls the init-operation which passes the GeneratorService' host address to be used for calling the GeneratorService' notify-operation upon finishing the calculation for a tile. The operation start starts the overall computation, and getImage retrieves the overall result which is composed of all sub-results obtained via the getTile-operation.

To run the distributed computations on a grid system with the Globus Toolkit as the middleware, additionally to implementing the operations described above, a specific middleware setup must be accomplished.

Here, the six steps of the middleware setup, shown in Fig. 2.8, are briefly listed and then explained in more detail for the presented case study.

1. Write the WSDL (Web Service Description Language) interface definitions. Since the required Web services operate on common data (the image to be generated, which Globus represents via a WS-resource) and communicate both synchronously (for invoking operations) and asynchronously (for notifying the caller about operation completions), the interfaces include non-standard WSDL elements which make their definition more complex than that of usual Web services.

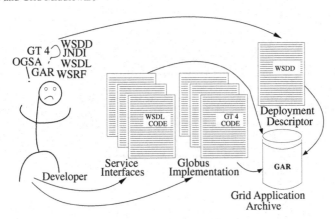

Figure 2.8 Programmer's Middleware Tasks Using Globus

2. Write a JNDI deployment file defining the Globus-typical factory classes used for remotely creating resources, called resource homes.
3. Develop service implementations (using the OGSA API in the Globus Toolkit).
4. Write the WSDD (Web Service Deployment Descriptor) configuration that defines where each service is implemented, e.g., in a Java class.
5. Package the Grid Application Archive (GAR) wherein the above files are combined together.
6. Deploy the services remotely, i.e., copy the GAR to the location specified in the WSDD file where the Globus container will find it.

Thus, the application programmer has to write by hand several bulky files in XML and other formats to work with the Globus middleware. This complex and laborious task makes the programmer look gloomy in Fig. 2.8.

The WSDL interface definitions for the GeneratorService (shortened) are shown in Fig. 2.9; for FractalService they are similar. For the sake of brevity, three of the four operation declarations were omitted, as were the associated parameter type and invocation message declarations (approximately 60–70 lines of XML-code). Note that all these pieces of code have to be written manually. Although it is possible to create WSDL-files from Java interfaces using commonly available Web services tools and extend them (to include the special namespace definitions and imports required by Globus for resource usage and notifications), this is quite error-prone. Thus, the Globus documentation explicitly advises to write the WSDL-code manually [cSC05].

Developing service implementations in OGSA is not easy, also because the documentation for programming Web services using WSRF is scarce and sometimes dated. Two of the few sources for information are the official Programmer's Tutorial [cSC05] from the Globus organization and the IBM Developer works Online Library [cIB07a]. Unfortunately, the illustrating examples in these documents only show how single services perform trivial tasks like basic arithmetics on integers.

```
<?xml version="1.0" encoding="UTF-8"?>
<wsdl:definitions name="GeneratorService"
                 targetNamespace="http://julia.core.chaos/Generator"
      xmlns:wsdlpp="http://www.globus.org/namespaces/2004/10/WSDLPreprocessor"
               ...    <!-- more namespace declarations        -->
                 xmlns="http://schemas.xmlsoap.org/wsdl/">
    <wsdl:import location="../../wsrf/properties/WS-ResourceProperties.wsdl"/>
               ...    <!-- more WSR-import statements          -->

    <wsdl:types>
       <schema targetNamespace="http://julia.core.chaos/Generator"
               xmlns="http://www.w3.org/2001/XML
         <import namespace="http://schemas.xmlsoap.org/soap/encoding/"/>
         <complexType name="ArrayOf_xsd_string">
           <complexContent>
             <restriction base="soapenc:Array">
               <attribute ref="soapenc:arrayType"
                          wsdl:arrayType="xsd:string[]"/>
             </restriction>
           </complexContent>
         </complexType>
         ...                    <!-- more parameter type declarations   -->
       </element>
      </schema>
    </wsdl:types>

    <wsdl:message name="configureRequest">
      <wsdl:part name="in0" type="impl:ArrayOf_xsd_string"/>
    </wsdl:message>
    <wsdl:message name="configureResponse">
      <part name="parameters" element="impl:void"/>
    </wsdl:message>       <!-- more message declarations       -->

    <wsdl:portType name="GeneratorPortType"
                 wsdlpp:extends="wsrpw:GetResourceProperty
                 wsrlw:ImmediateResourceTermination"
                 wsrp:ResourceProperties="tns:MasterResourceProperties">
      <wsdl:operation name="configure" parameterOrder="in0">
        <wsdl:input message="impl:configureRequest"/>
        <wsdl:output message="impl:configureResponse"/>
      </wsdl:operation>
      ...                    <!-- more operation declarations        -->
    </wsdl:portType>
</wsdl:definitions>
```

Figure 2.9 Simplified WSDL Interface for the GeneratorService

In contrast, even our relatively simple fractal image application uses multiple Web services and resources: The `GeneratorService` where the workspace (an initially empty matrix, wherein the image is generated) is represented using resources (one per image instance) and several instances of the `FractalService` that are processing segments of the image (each represented using a resource of its own). So far, there is still no documentation available which describes WSRF applications using multiple interconnected Web services and resources.

The archive (GAR) creation can be done using the jar-Tool which is a part of the Standard Java SDK. However, the contents need to be arranged manually beforehand: the developer must set up a directory structure containing multiple records of binaries and configuration files, which can be quite complex, especially for multiple, interrelated Web services. In the case of the fractal image application, this includes of course all binaries performing the required service activity. This results in altogether 39 class files that must be arranged according to the 5-level package hierarchy of the application. The 5-level hierarchy results from the structuring of the application classes into utilities for creating the graphics, utilities for the computation, internally used data record declarations, and service-related functionality such as resource homes, service interfaces and their implementations. These 39 files contain approximately 500 lines of hand-written Java code (including, besides the

Julia Set computation code, some utilities for computations on complex numbers and image processing, which were developed for the application) and 4000 lines of Java code which the Globus Toolkit generated automatically from the user-defined WSDL files (holding the Java files for the service interfaces and factories).

Furthermore, the required middleware setup includes a schema directory containing a file with namespace-to-package mappings, two deployment descriptor files for the two different services types in the application and, for each service, a subdirectory containing the hand-written WSDL, as well as additional WSDL-files defining the resource factory interfaces, service binding files and XML schema definition files that are automatically generated by the Globus service deployment tools. For locating base classes of, e.g., resource homes, the deployment descriptor files refer explicitly to subpackages of the OGSA packages contained in the GT 4 implementation, whose structure will be probably subject to change in the future releases of the Globus Toolkit.

```
<?xml version="1.0"?>
<deployment name="defaultServerConfig" xmlns="http://xml.apache.org/axis/wsdd/"
        xmlns:java="http://xml.apache.org/axis/wsdd/providers/java">

    <service name="chaos/core/julia/GeneratorFactoryService"
            provider="Handler" style="wrapped">
            <parameter name="operationProviders"
                    value="org.globus.ogsa.impl.ogsi.NotificationSourceProvider"/>
            <parameter name="instance-baseClassName"
                    value="chaos.core.julia.impl.GeneratorImpl"/>
            ...      <!-- more service parameters      -->
    </service>

    <service name="chaos/core/julia/FractalFactoryService"
            provider="Handler" style="wrapped">
            <parameter name="instance-baseClassName"
                    value="chaos.core.julia.impl.FractalImpl"/>
            ...      <!-- more service parameters      -->
    </service>
</deployment>
```

Figure 2.10 Simplified WSDD for the Fractal Application

According to the current technology, the WSDD configuration is also supposed to be written manually. Its simplified version for the case study is shown in Fig. 2.10 (here, additional 45 lines of declarations are omitted for brevity). The contents, format and syntax of these files depend on the Globus Toolkit version used. Unfortunately, no tools are available for checking the correctness of WSDL and WSDD-files for Globus. Thus, possible errors in the extensive Globus service configuration code are difficult to detect.

Summarizing, even the comparatively simple case study, whose parallel structure is quite straightforward (*embarrassingly parallel*), requires the application programmer to have a lot of specific, technical knowledge in the field of the grid middleware. In a hardwired implementation, the contents of the configuration files are application-dependent, which makes the task of programming even harder: while the application evolves, the programmer will have to rewrite the contents again and again. A change of the used version of middleware may also require rewriting all or some of the files. A possible change of the middleware system may have even more serious consequences: e.g., a transition from Globus to Unicore would require the programmer to completely redesign the application.

The main problem with this programming approach is not only that the user has to arrange numerous low-level, system-specific files; even more frustrating is that all these details have very little to do with the application itself. Current middleware and also systems anticipated for the next few years do not liberate the application programmer from grid technicalities. The user is still distracted from his proper business: improving the application, finding suitable parallelization strategies, etc.

2.2.2 Bridging Middleware and Application with HOCs

In this section, the HOC approach from Section 2.1 is reviewed. While in Section 2.1 the focus was on the Java implementation, this section shows how the generic configuration of the Farm-HOC frees the application programmer from writing a new middleware configuration for every new application.

The low-level nature of current grid middleware and its complexity are rooted in its inherently complicated task: mediating between two very different levels of abstraction. On the one hand, the grid possesses a complex, heterogeneous and dynamical physical structure. On the other hand, grid applications exemplify a tremendous richness of algorithmic structures, programming paradigms, programming languages and modes of parallel computation and communication. While middleware systems like Globus efficiently capture the low-level details of grid behavior, their interface to application programmers needs to be improved by increasing the provided level of abstraction.

The configuration files for the case study (Fig. 2.9 and Fig. 2.10) hold application-specific information, e.g., the name of the FractalService and its output format (a bitmap image). So, here the features of the application are "hardwired" into the Web services. This means that for any different application, all the laborious implementation and configuration steps must be accomplished anew, even if the next application adheres to the farm pattern again. Using HOCs (as introduced in Section 2.1) is a much more generic approach than using such hardwired services.

The key concept of HOCs is to provide implementations of generic, recurring patterns of parallel behavior for remote selection, customization (via the code parameters) and combination. "Recurring" means that the pattern is used again and again in different applications. "Generic" means that a component is independent of a particular application, but can be customized to a particular case using appropriate parameters. "Combination" means that different HOCs can be composed together, e.g., a farm of pipelines can be built this way.

The process of grid application programming using HOCs is shown in Fig. 2.11. The programming and middleware setup tasks are divided between two groups of programmers: grid system experts and application programmers. While the former prepare the necessary implementations and middleware configuration for HOCs, the latter develop applications using pre-implemented HOCs. HOCs free the application

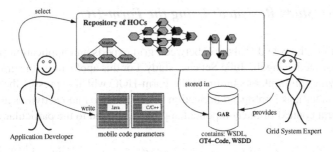

Figure 2.11 Using HOCs: The Idea

programmer from low-level arrangements: he can concentrate on the application itself, that is why he looks more happy in Fig. 2.11 than he looked in Fig. 2.8. The person "with a hat" in the figure, is the grid system expert. It is the grid system expert who in the HOC approach should free the application programmer from much of the burden related to the necessary middleware setup.

Note that the pre-configuration does not result in a fixed execution topology. HOCs are typically not residing on the host providing the Web service to access them and the locations where the actual computations run are determined during scheduling, which is done at runtime (either via calling `configuregrid` or via an automatic scheduler as shown in Chapter 5).

For the application programmer, program development proceeds as follows:

- Select suitable HOCs for the application from the repository of HOCs.
- Express the application by customizing HOCs with application-specific code parameters expressed, e.g., as Java code (and, if required by the application, composing multiple HOCs together).
- Rely on the pre-packaged implementation of the selected HOCs, available in the repository.

The grid system expert develops the grid implementation of each HOC, including the necessary middleware setup. Owing to his detailed knowledge of the particular grid system and its middleware, the grid system expert can typically accomplish a higher-quality implementation than the application programmer. The expert prepares efficient parallel and/or remote implementations and middleware setups of HOCs and bundles them in grid archives, which are deployed to the servers of the grid. Note that the task of the grid system expert in arranging middleware in Fig. 2.11 is even simpler than it was for the application programmer in Fig. 2.8: whereas the expert prepares the setup once for each component, the programmer had to do it again and again for each application.

2.2.3 Case Study Revisited: Using the Farm-HOC

In this section, the role distribution between grid experts and application programmers is illustrated using the Julia Set case study. For this application, the grid system expert provides an implementation of the Farm-HOC which expresses the farm algorithmic pattern, as introduced in Section 2.1.2. The two main activities are shown in the general context of the component approach, applied to the particular example application:

- First, it is demonstrated how the Farm-HOC is implemented by the grid system expert, together with the necessary middleware setup.
- Second, it is demonstrated how the Farm-HOC can be used for programming the case study of fractal image computation.

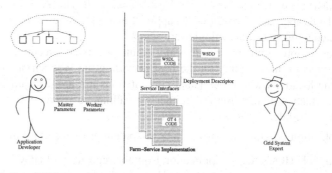

Figure 2.12 Farm-HOC: Developer's and Expert's View

Figure 2.12 shows the distribution of roles between the application programmer and the grid expert in case of the Farm-HOC. Both persons have a similar understanding of the farm pattern as a structure with one master and several workers. The difference is that while the grid system expert concentrates on the generic control structure of the farm pattern, regardless of the particular activities of the master and the workers, the application programmer takes the farm control structure for granted and pays his attention rather to the contents of the master and the workers, i.e., their particular instances.

Summarizing, the roles between the grid system expert and the application programmer in the case of Farm-HOC are divided as follows:

- The grid system expert develops a grid implementation of the generic Farm-HOC, i.e., of the farm control structure in Fig. 2.12, including both parallelization and the middleware setup.
- The application programmer uses the Farm-HOC by providing two customizing codes for its parameters: `master` and `worker` which, in the farm control structure, fill in the missing pieces (in Fig. 2.12, the empty boxes) such that the first parameter defines the master (the big box) and the following parameter defines

the worker code (copied, for each of the smaller boxes, since each worker runs the same code).

An important question is: how big is the fraction of the effort on the parallel implementation and necessary middleware configuration of a particular HOC that can be undertaken by the grid system expert in advance, i.e., independently of a particular application? The answer for the Farm-HOC is as follows:

- Implementations of Web services (farm's master and worker) are provided by the grid system expert. Different parallelization strategies can be used, depending on the architecture of the grid hosts involved.
- By choosing generic parameter types, it is possible to free the programmer from the tedious configuration work in Globus, i.e., from writing WSDL/WSDD and packaging GARs.

```xml
<?xml version="1.0" encoding="UTF-8"?>
<?xml version="1.0" encoding="UTF-8"?>
<wsdl:definitions name="MasterService"
                  targetNamespace="http://org.gridhocs/Master"
   xmlns:wsdlpp="http://www.globus.org/namespaces/2004/10/WSDLPreprocessor"
               ...   <!-- more namespace declarations          -->
                  xmlns="http://schemas.xmlsoap.org/wsdl/">
   <wsdl:import location="schema/ogsi/ogsi.gwsdl"
       namespace="http://www.gridforum.org/namespaces/2003/03/OGSI"/>

   <wsdl:types>
     <schema targetNamespace="http://org.gridhocs/Master"
             xmlns="http://www.w3.org/2001/XML
       <import namespace="http://schemas.xmlsoap.org/soap/encoding/"/>
       <complexType name="ArrayOf_xsd_double">
     <complexContent>
      <restriction base="soapenc:Array">
        <attribute ref="soapenc:arrayType" wsdl:arrayType="xsd:double[]"/>
      </restriction>
     </complexContent>
    </complexType>
       ...              <!-- more parameter type declarations    -->
      </element>
     </schema>
   </wsdl:types>

   <wsdl:message name="configureRequest">
     <wsdl:part name="in0" type="impl:ArrayOf_xsd_string"/>
   </wsdl:message>
   <wsdl:message name="configureResponse">
     <part name="parameters" element="impl:void"/>
   </wsdl:message>       <!-- more message declarations          -->

   <wsdl:portType name="MasterPortType"
               wsdlpp:extends="wsrpw:GetResourceProperty
               wsrlw:ImmediateResourceTermination"
               wsrp:ResourceProperties="tns:MasterResourceProperties">
     <wsdl:operation name="configure" parameterOrder="in0">
       <wsdl:input message="impl:configureRequest"/>
       <wsdl:output message="impl:configureResponse"/>
     </wsdl:operation>
       ...              <!-- more operation declarations         -->
   </wsdl:portType>
</wsdl:definitions>
```

Figure 2.13 Generic WSDL Interface for the Farm-HOC Master

This is illustrated by the WSDL file for the Farm-HOC `MasterService` in Fig. 2.13. It is very similar to the `GeneratorService` in Fig. 2.9. For the `MasterService`, there is also a `configure`, a `start`, a `notify` and (corresponding to `getImage`) a `getResults` operation. The only difference is that the `MasterService` uses generic types for all parameters and results (i.e., `xsd:double[]`), while in the `GeneratorService`, the same data was declared in an application specific manner (as a pair of coordinates in the complex number plane and the maximum iteration depth).

- A generic WSDD can also be provided, defining which class implements which HOC. The difference between the generic version and the deployment descriptor shown in Fig. 2.10 is that the service elements of the generic WSDD hold references to the `MasterService` and the `WorkerService` implementation (instead of the particular `GeneratorService` and `FractalServices`).

- Once the configuration files have been arranged and packaged into a GAR, the archive can be deployed by copying it to the target host and registered by a directory service, which is also done by the grid system expert. Thereupon the Farm-HOC is available to the application programmer via a Web service. The application programmer only needs to customize the HOC with an application-specific code.

Summarizing, a major part of both parallel service implementations and the middleware setup for the Farm-HOC can be accomplished in an application-independent manner. In the case of the fractal image application, the application-specific code (master/worker parameters and the complete client code) constitute about 21% of the total lines of code.

2.2.4 Performance Experiments on a Wide-Area Testbed

Table 2.1 shows the results of some tests conducted with the Farm-HOC. The experimental grid testbed consists of one host in Münster running the master implementation, and up to three remote multiprocessor hosts in Berlin (at a distance of approx. 500 km), each running multiple parallel workers. The underlying TCP/IP network has the bandwidth of 1 MB/sec and the latency of 25 ms.

Table 2.1 Performance Measured for the Farm-HOC

1 remote server (4 processors)	2 remote servers 4 + 8 (processors)	3 remote servers (4 + 8 + 12 processors)
198,212 sec	128,165 sec	48,377 sec

The server in Münster was a Linux PC (Pentium 4 running at 2.6 GHz) and the remote servers were SunFire multiprocessors with 4, 8, and 12 processors (all running Solaris 10).

For Julia Sets, all tasks have different time costs and the compute hosts have different computing power, so the scalability behavior is not regular. Nevertheless, the results show that the application does scale with increasing the number of processors. The variations in multiple measurements were low. The sequential time of a local evaluation on the PC was more than five times larger than using only the slowest remote server with 4 processors.

The application chosen as the first example (Julia Set computation) can be easily tuned to clearly manifest the scaling behavior of the Farm-HOC: since the user can decide after how many iterations a complex number is classified as the starting point for either a divergent or a convergent sequence, the user can simulate the behavior of a very complex computation by increasing this limit. In the reported measurements, the iteration limit was set to 500, leading to a significant evidence of scalability (but also to relatively high time costs for this simple application).

Another result of the experiments was that transferring the result via SOAP takes much time (about 60 sec), due to the complexity of the SOAP encoding. A possible optimization, which does not require the use of a different middleware, would be to exploit GridFTP in Globus for transferring the results. The Java CoG-Kit [bL+01] provides an API that simplifies the use of GridFTP in Java-based programs and might be used for this purpose.

2.2.5 HOCs and Hand-Written Code: A Performance Comparison

In Section 2.2.4, some performance experiments with the Julia Set case study were reported. The alternative, hardwired implementation, shown in Section 2.2.1 was used to conduct a performance comparison with the Farm-HOC.

The studied question was: What is the time overhead incurred by using HOCs as an additional level of abstraction between the grid middleware and application programmers? The runtimes of the two versions of the fractal image application on the same grid testbed were compared: (1) the hardwired version, with the middleware configuration and parallel implementation written manually, and (2) the version with the Farm-HOC where the middleware configuration and parallel implementation is pre-packaged, and the application is expressed via the code parameters.

Interestingly, no measurable difference in the runtime of these versions could be observed (absolute values were shown in Table 2.1 in Section 2.2.4). In repeated measurements the execution times were almost equal with variations of less than 2%. The measured times rank the pure calculation time, i.e., the time that elapsed between the invocation of the farm master's `start`-operation until the master server received the last of all `notify` calls from the employed workers.

The almost equal results reflect that for an object at runtime, it does not make a difference if its defining class was loaded from the local hard disc or transferred from a remote computer. Of course, there is a slight difference in the way parameters are passed to both versions of the services at runtime: `GeneratorService` and `FractalService` work with dedicated variables, while `MasterService` and `WorkerService` use generic number arrays and convert them where necessary. However, a time difference for this simple conversion was not measurable in the experiments.

In the version using the Farm-HOC, additional initialisation time is required for transferring the code parameters: The master code must be transferred to the MasterService-host and the worker code parameter to all WorkerService-hosts. Transferring the worker parameter with a size of 3857 bytes took 90 ms at average for each worker and transferring the master with a size of 5843 bytes took 122 ms at average. For 12 workers, this results in a time of approximately 1.2 sec, which can be disregarded, because this is a nonrecurring setup time that is spent only once when a HOC is used for the very first time. Unfortunately, in the most simple possible setup with only one grid server hosting the Code Service and running the parallel HOC implementation, the submission must be done again, when a code parameter changes (in Section 3.1.2, it is shown how code parameters can be stored in dedicated databases, including object-relational ones, which reduces the initialization time for HOCs).

The network bandwidth during the measurements was approximately 1 MB/sec and latency was less than 25 ms, which implies the expected transmission time of less than 1 sec for 12 workers when using plain TCP/IP communication. The measured additional overhead of slightly more than 0.2 sec is due to the employed SOAP encoding mechanism, which causes an additional culmination of descriptive data.

An almost linear speedup of the case study application can be achieved when the plain farm implementation is extended by a dynamic load balancing strategy, such as a task redistribution at runtime via a task-stealing mechanism [aSB99]. Advanced scheduling and load-balancing strategies for HOCs are discussed in Chapter 5.

2.3 APIs for Grid Application Programming with HOCs

Convenient APIs for programming new HOCs and applications based on them on a high level of abstraction were developed. There is a Client API and a server-side Service API [bA+04] and both APIs are available online for downloading [cD+06].

The focus of this section is on the design of the server-sided framework and API used for implementing HOCs.

As discussed in Section 2.2, the process of grid application programming using HOCs is split between two different communities of programmers. These are: (1) grid system experts who design and implement HOCs including the setup work required for the target platform, and (2) application programmers who use HOCs to compose applications by parameterizing (and combining) the available HOCs.

Figure 2.14 shows the structure of the developer APIs (comprising the *Client API* and the server-side *Service API*) for HOC programming.

Application programmers work with the simple and well-documented Client API for passing code parameters, accessing services and retrieving results from remote hosts (how to use this API for invoking HOCs was already shown in Fig. 2.5).

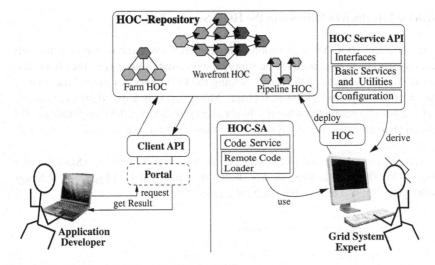

Figure 2.14 Developer Roles and the APIs for HOCs

For convenience reasons, there is also a portal that allows users to submit code parameters written in different languages to HOCs via a Web browser, including an interactive mechanism for handling errors in the uploaded code at runtime (shown in Section 3.2.4). The parameters of a HOC are application-specific codes provided by users who are specialists in a specific scientific domain, e.g., biochemists, physicists or seismologists.

The right part of Fig. 2.14 shows that the grid system expert works with an API that defines basic services and utilities, common interfaces and a standard configuration to derive new HOCs. The Code Service and the Remote Code Loader are provided by the HOC-SA [bDG04]. The Remote Code Loader (explained in more detail in Chapter 3) frees the grid system expert from dealing with raw binary data when creating instances of classes, defined in code parameters uploaded from a client. The utilities provided by the HOC Service API comprise a library of recurrently used data types (number sets, matrices, vectors, etc.) and procedures (splitting, transposition, etc.) that help to implement parallel numerical applications with little effort (see Section 2.4.2 for an example), as well as an efficient implementation of group communication (explained below). Grid system experts must be capable of writing efficient parallel code for the target grid architecture and the employed middleware system. Although generic definitions of interfaces and classes for HOC development are provided by the HOC Service API, a grid system expert who develops new HOCs is still required to extend the basic definitions and arrange all the configuration files appropriately for each HOC (as explained in Section 2.2).

This section shows how to use the HOC Service API and the HOC-SA, while Chapter 3 explains how the HOC-SA works.

Generic Interface Definitions in the HOC Service API

What the Farm-HOC has in common with other HOCs is that it is implemented only partially, and therefore, needs to be supplied with a data source (i.e., the client must specify the IP-Address of the server hosting the Code Service) to load the missing pieces of the implementation at runtime. Moreover, it is a distributed component that needs to be instructed what machines to employ for parallel computations. The special characteristics of this HOC are, that there are two code parameters, `Master` and `Worker`.

This design is reflected in the service interface represented in WSDL notation (Fig. 2.15). The code is considerably shorter than the WSDL in Fig. 2.13, although both figures show a remote interface for a farm.

```
<wsdl:portType name="FarmPortType"
               wsdlpp:extends="hocsa:GridHOC">
  <operation name="selectMaster">
    <input message="tns:SelectMasterInputMessage"/>
    <output message="tns:SelectMasterOutputMessage"/>
    <fault name="Fault" message="hocsa:FaultMsg"/>
  </operation>
  <!-- selectWorker analogously -->
</wsdl:portType>
```

Figure 2.15 PortType Definition of the Farm Service

The compact notation in Fig. 2.15 is facilitated by the generic base definitions in the HOC Service API. Only the special characteristics of the farm have to be specified: two `operation`-elements for selecting the `Master` and the `Worker` code unit by passing keys of the `xsd:string` type. While the interface in Fig. 2.13 was directly derived from `wsrpw:GetResourceProperty` (required for data sharing among the operations), the interface in Fig. 2.15 is derived from `hocsa:GridHOC` inheriting the `wsrpw:GetResourceProperty` parent interface and many more definitions.

An extract of the basic definitions of `hocsa:GridHOC` is shown in Fig. 2.16. Web services are a non-object-oriented technology, and therefore, inheritance is not supported by standard Web service runtime environments. However, new HOCs that are developed using the Service API are derived from `hocsa:GridHOC` as shown in Fig. 2.15. The HOC Service API makes use of the `extends` attribute (bound to the Globus-specific namespace `wsdlpp`) which allows to use simple and multiple inheritance in the context of WSRF [cOA04]. The definitions inherited by using the `hocsa:GridHOC` base interface (defined in the Service API) typically make up the biggest part of the service interface of a HOC. This file comprises the definitions which are necessary and common for all HOCs. Here, the Service API makes use of the support for multiple inheritance in Globus: A HOC may be both a WS-N [cOA06] producer and a consumer of notifications (i.e., messages sent to or received over the Internet asynchronously).

```
<wsdl:portType name="GridHOC"
   wsdlpp:extends="wsrpw:GetResourceProperty"
   wsntw:NotificationConsumer wsntw:NotificationProducer"">
 <operation name="setDatasource">
  <!-- configuration msg. declarations... -->
 </operation>
 <operation name="getDatasource">... </operation>
 <operation name="configureGrid">... </operation>
 <operation name="getMachine">... </operation>
 <operation name="compute">... </operation>
 <operation name="getResult">... </operation>
</wsdl:portType>
```

Figure 2.16 Basic Interface of a HOC (shortened)

Generic Class Definitions in the HOC Service API

The HOC Service API includes shell-scripts to bundle and deploy new HOCs (derived from `hocsa:GridHOC`) in GAR files automatically.

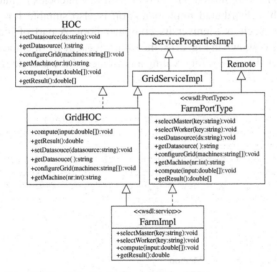

Figure 2.17 The Farm-HOC and Its Parent Classes in the API

This includes the generation of the stub code that interfaces between the service implementation and the middleware. So, what remains to be done for the grid system expert when a new HOC is developed is to provide service implementations. For the Farm-HOC, this is the `FarmImpl`-class shown at the bottom of the class diagram in Fig. 2.17.

An important feature of HOCs is that there may exist multiple implementations for one and the same HOC, each tuned to a particular hosting environment. For the

Farm-HOC, one might think of an implementation that consists of only one `Farm`-service running the `Master`-unit, and several threads, all running the `Worker`-unit on the same server. This solution is preferable if the service host is a large multiprocessor machine which can run a master and several workers in parallel.

```
1: public void selectMaster(String masterKey)
2: throws RemoteException  {
3:   codeService.setDatasource(clientCodebase);
4:   Class masterClass =
5:     remoteCodeLoader.load(masterKey);
6:   this.master = (Master)masterClass.newInstance( );   }
```

Figure 2.18 The `selectMaster` Method

The Farm-HOC implementation which was used for the fractal image case study in Section 2.1.5 follows a multi-service approach where each worker is accessed via a Web service of its own, running on a different host than the `Farm`-service. This distributed implementation is most suitable e.g., if both the `Farm`-service and the `Worker`-services are hosted on a cluster with low latencies of inter-node communication. Such a distributed implementation is also suitable for a network of idle workstations, such as a university computer pool. A part of the code of the `FarmImpl`-class is shown in Fig. 2.18: the `selectMaster`-method using the Remote Code Loader (lines 4–5) for creating an instance of the user-defined `Master`-implementation.

```
1: public void selectWorker(String workerKey)
2: throws RemoteException   {
3:   GridDataServiceFactory gdsf =
4:     ServiceFetcher.getFactory(getDatasource( ));
5:   GridDataService gds = factory.createGridDataService( );
6:   SQLQuery query = new SQLQuery( "select codeUnit from"
7:     + " mobileCode where unitID='" + workerKey + "'" );
8:   Response response = gds.perform(query);
9:   workerCode = response.getAsString( );
10: gds.destroy( );   }
```

Figure 2.19 The `selectWorker` Method

The Service API also allows the grid system expert to bypass the Remote Code Loader and interact with the data sources directly using the OGSA-DAI [cUD07] framework. An example of this can be seen in the `selectWorker`-method in Fig. 2.19, where a Grid Data Service (GDS [cUD07]) is created which is connected to a relational database. By invoking the `perform`-operation of the GDS (line 8), the worker code is retrieved via an SQL Query and stored intermediately. The reason for directly connecting to a GDS in the `selectWorker`-method of the example Farm-HOC implementation is that the `Farm`-master does not need an instance of the class defined by the worker parameter. Here, the binary code of the user-defined `Worker` class is loaded without instantiating it for passing it through to the worker hosts.

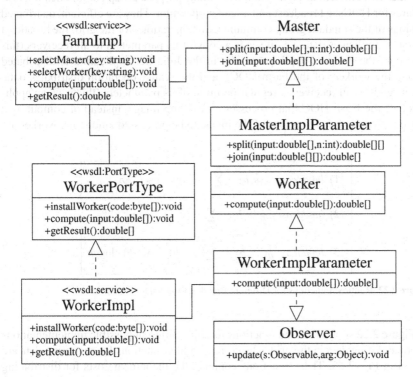

Figure 2.20 Farm-HOC Evaluation by Delegation

The class implementing the `Worker`-service delegates calls to `WorkerImplParameter` for processing single subsets (Fig. 2.20). The `WorkerImplParameter` internally forks threads of the `Evaluator`-type (shown in Fig. 2.21), that process input items in the background and sends back notifications (line 8) whenever one of these threads is finished.

```
1: class Evaluator extends Observable implements Runnable  {
2:     ... // attribute definitions
3:     public Evaluator(double[] input, Worker worker)  {
4:         ... /* store arguments in
5:                 private attributes */  }
6:     public void run( )  {
7:         result = worker.process(input);
8:         setChanged( );  notifyObservers( );  }
9:     public double[] getResult( )  {
10:        return result;  }  }
```

Figure 2.21 Evaluator Thread Running in the Worker Service

The Group Communication API for HOCs

Besides generic class and interface definitions, the HOC Service API (Fig. 2.14) provides two efficient implementations of multi-purpose group communication procedures for HOCs: a broadcast and a scatter operation. They are efficient, distributed versions of the standard group communication operations for the grid: the broadcast operation replicates data, i.e., it sends a copy of its parameters to all receivers (this way, e.g., the iteration-depth parameter in the Julia Set application is distributed among the workers of the Farm-HOC), and the scatter operation distributes data, i.e., it sends to all receivers a regular partition of its parameters (this way, an application of the Farm-HOC that processes an existing image, instead of computing a new one, will distribute regions of the image to be processed among the workers).

$$(1)\ broadcast_{seq.}(n,b) = \sum_{i=1}^{n}(v+b) = (v+b)*n$$

$$(2)\ broadcast_{multi.}(n,b) = \sum_{i=1}^{n}b+v = v+b*n$$

$$(3)\ broadcast_{orth.}(n,b) = \sum_{i=1}^{\log_2 n}(v+b) = (v+b)*\log_2 n$$

Figure 2.22 Runtime Functions for Different Broadcast Implementations

Figure 2.22 shows runtime functions for three different possible implementations of broadcasting data in the grid. The corresponding functions for scattering data are shown in Fig. 2.23. These functions compute the theoretical costs for distributing some data to n grid nodes, supposed the costs for a single transmission are b and establishing any possible network connection has always the same cost v.

$$(4)\ scatter_{seq.}(n,b) = \sum_{i=1}^{n}\left(t+v+\frac{b}{n}\right) = (t+v)*n+b$$

$$(5)\ scatter_{multi.}(n,b) = \sum_{i=1}^{n}\left(t+\frac{b}{n}\right)+v = t*n+v+b$$

$$(6)\ scatter_{orth.}(n,b) = \sum_{i=1}^{\log_2 n}\left(t+v+\frac{b}{n}\right) = (t+v)*\log_2 n$$

Figure 2.23 Runtime Functions for Different Scatter Implementations

In the most naive implementation which distributes the data among all nodes in turn (formulas (1) and (4)), whereby (4) has an additional summand t for partitioning the data, as required for scattering), all costs are simply summed up. Formulas (2) and (5) refer to the implementation which starts all sending processes at once, each in a thread of its own, such that v is no more a part of the sum. The best solution for sending large data masses is probably, a tree-structured distribution which reduces the upper limit of the sum to $log_2 n$ as shown in formulas (3) and (6), since the number of sending processes is doubled on each stage. In practice, only the distribution of very little data among a very small number of nodes can be done faster with a sequential or multithreaded group communication operation, due to the overhead of the tree-structured implementation which, theoretically, is always superior.

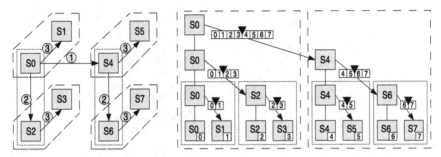

Figure 2.24 Orthogonal Group Broadcasting/Scattering in the grid

There are multiple versions of the HOC group communication operations in the API for broadcasting and scattering different data types in the grid. All HOC group communication operations follow a structure that is even more efficient to a simple tree. The implementation is based on *orthogonal communication patterns* [bR⁺01], which have been proven to be a very efficient variant of implementing MPI-based group operations on local clusters. Therefore, formulas (3) and (6) are labeled *orth.* in the figures. Figure 2.24 shows how the implementation of the orthogonal communication patterns for HOCs works for eight grid nodes. If one follows the arrows, it can be seen how the data is distributed.

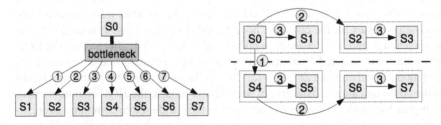

Figure 2.25 Different Group Communication Structures

Figure 2.25 compares linear group communication to the orthogonal implementation of broadcasting data for eight grid nodes (i.e., networked computers). On the left, it shows the linear group communication starting from node "S0" to the nodes "S1-S7", leading to a bottleneck on the S0-link. The orthogonal implementation in the HOC Service API (Fig. 2.25, right) avoids the bottleneck by continuously dividing the available nodes into a hierarchy of groups and subgroups until the deepest subgroups contain only two or less nodes. These groups can be graphically arranged in rows and columns, therefore the name orthogonal communication.

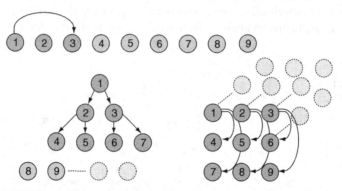

Figure 2.26 HOC Group Communication

Figure 2.26 shows the difference between the HOC group communication and a simple tree. Here, for a network of nine grid nodes, the number of nodes that have been supplied with new data after three communication steps are highlighted. The top row in the figure shows the effect of linear communication again: only three nodes received the data (1 ... 3), while six are still waiting. The bottom part of the figure compares a simple tree (left) and the orthogonal HOC communication structure (right). After the second step, the HOC communication operation has outperformed the tree version: all nine nodes have received the data. In the first step, node 1 supplies nodes 2 and 3 with the data, since they are in the same row. The original tree operation has the same effect, since nodes 2 and 3 are direct children of node 1. But in the second step, the tree operation continues with sending the data from node 2 to nodes 4 and 5, and, simultaneously, from node 3 to nodes 6 and seven, while node 1 stays idle. The HOC operation in contrast, runs three simultaneous sending processes for the columns: node 1 sends to nodes 4 and 7, node 2 sends to nodes 5 and 8, and, node 3 sends to node 6 and 9. In general, after step n, the HOC operations have updated 3^n nodes with data, while a standard tree communication operation can only update $\Sigma_{k=0}^{n} 2^n = 2^{n+1} - 1$ at the same time. Due to the structuring of the communication in row groups and column groups, the HOC operations have no higher bandwidth demands as compared to the standard tree communication, although there are more concurrent sending processes. As observed in Fig. 2.26, these processes use different network links: in the second step, there are still only two receivers per message, as in the original tree.

Experimental Evaluation of HOC Group Communication

Commonly used grid programming libraries, e.g., ProActive [bB⁺02] offer group communication operations, similar to the broadcast and the scatter operation in the HOC Service API. Contrary to the HOC communication operations, these libraries typically implement group communication following a linear structure which leads to performance issues with a growing number of communicated data and participating grid nodes. To show the benefits of the HOC group communication operations, the broadcast is compared to the corresponding group communication operation in the ProActive library. Note that the experiments were conducted using a ProActive version that was available in 2006. This library can be seen as a representative of any grid programming library where data is communicated linearly. In the most recent ProActive versions, the group communication has been optimized, adopting a structure similar to the implementation in the HOC Service API.

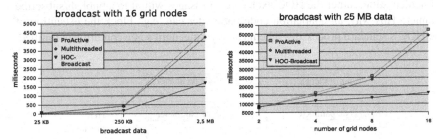

Figure 2.27 Performance Results for the Broadcast

Figure 2.27 shows the results of experiments with three different broadcast implementations in a local-area network with a data throughput between 3 and 4 MBit/s. To avoid I/O-delays, a thread pool was used in a variant of linear communication, which starts multiple linear sending processes at once. Therefore, these linear operations are labelled "Multithreaded". The left part of the figure shows a test involving 16 grid nodes and a growing amount of data. The right part of Fig. 2.27 shows the same group communication, but with a fixed amount of data (25 MB) and a growing number of involved grid nodes. In both cases the advantage of the HOC broadcast implementation can be observed in the diagram. Instead of exponentially growing communication times, there is linear growth with every doubling of involved grid nodes for the scatter and broadcast operation in the HOC Service API.

2.4 Adaptability of HOCs

This final introductory section on HOCs shows that code parameters can not only be used for component customization but also for component adaptation. While a *customization* sets a specific operation within the parallel processing schema of a component, an *adaptation* changes a component's parallel behavior [bG⁺06].

Contrary to customization, adaptation requires from the programmer to know and understand the HOC's parallel processing schema, but it still frees the programmer from a lot of implementation and setup work: if a programmer requires a particular component that is not yet available in the server-side component framework, but a HOC with a similar functionality exists, the programmer can adapt the existing HOC instead of implementing and configuring a completely new component from scratch.

2.4.1 Code Parameters for Adaptation

In contrast to a customizing code parameter, which is applied within the execution of the HOC's schema, a code parameter specifying an adaptation runs in parallel to the execution of the HOC. There is no fixed position for the adaptation code in the HOC implementation; rather the HOC exchanges messages with it in a publish/subscribe manner. This way, an adaptation code parameter can, e.g., block the execution of the HOC's standard processing schema at any time, until some condition is fulfilled.

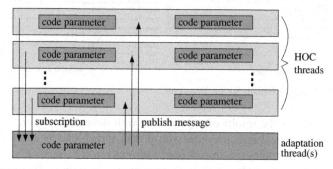

Figure 2.28 The Messaging Schema for HOC Adaptations

Figure 2.28 depicts the thread-messaging model for adapting HOCs. Initially, all the threads executed by the HOC perform a subscription by sending a message to the adaptation code parameter which is provided by the user. The adaptation thread is fully defined by this code parameter, contrary to the HOC threads which use code parameters for executing application-specific operations within their inherent schemata. The adaptation code parameter executes its operations concurrently to the standard schema (in terms of Java, it implements the `Runnable` interface); the threads running inside the HOC communicate with the adaptation code by publishing data, i.e., by sending messages to all the threads registered in the subscription. This messaging model allows adaptation code parameters to be exchanged or even omitted, while every HOC application requires the user to specify all customizing code parameters.

When only customizing code parameters are used in an application, HOCs abstract over the mechanisms that are employed to execute their inherent schemata in parallel on the grid. Code parameters expressing adaptations are optional and allow to intervene in the internal mechanisms of a HOC in order to alter its processing schema, e.g., from farm towards wavefront, as explained in Section 2.4.2.

The HOC implementation design can be viewed as a general method for making components adaptable. The two most notable characteristics of the presented implementation are the following:

(1) using HOCs, adaptation code is placed within one or multiple threads which are independent of the threads executed by the HOC, i.e., the original component code remains unchanged, and

(2) an adaptation code parameter is connected to the HOC using only message exchange.

This design has the following advantageous properties:

- Adaptation code is clearly separated not only from the component implementation code, but also from the obligatory, customizing code parameters. Thus, the data dependences, which are specific to a particular parallel algorithm for processing an application, are completely hidden in the adaption code. When a new algorithm with new dependences is implemented, the customization parameters can still be written as if this algorithm introduced no new data dependences.

- There is no fixed correlation between the number of obligatory customizing code parameters and the number of optional adaption parameters, but there can be an arbitrary number of adaptations. Due to the messaging model explained above, adaptation code parameters can easily be exchanged. It is not required that adaptation code parameters have a specific signature: Any thread can publish messages for delivery to other code that provides the publisher with an appropriate interface for receiving messages. Thus, adaptations can also adapt other adaptations and so on.

- The presented implementation offers location independence which is especially important in dynamic networked infrastructures like the grid. In the Farm-HOC, e.g., the data to be processed can be placed locally on the machine running the master or they can be distributed among several remote servers. In contrast to coupling the adaptation code to the `Worker` code, which would be a consequence of placing it inside the same class, adaptations are not restricted to affecting only the remote hosts executing the `Worker` code, but can also have an impact on the host running the `Master` code. In the following case study, this feature is used to efficiently optimize the scheduling behavior with respect to data locality: processing a certain amount of data locally on the master host significantly increases the efficiency of the computations.

2.4.2 Case Study: From Farm to Wavefront

The case study in this section is one of the fundamental algorithms in bioinformatics – the computation of *distances* between DNA sequences, i.e., finding the minimum number of insertion, deletion or substitution operations needed to transform one sequence into another. Sequences are encoded using the nucleotide alphabet $\{A, C, G, T\}$, where each letter stands for one of the nucleotide types [aB$^+$91].

The distance, which is the total number of the required transformations, quantifies the similarity of sequences [bLe66] and is often called *global alignment* [bH$^+$90]. Mathematically, global alignment can be expressed using a so-called *similarity matrix* S, whose elements $s_{i,j}$ are defined as follows:

$$s_{i,j} := max\left(s_{i,j-1} + plt, s_{i-1,j-1} + \delta(i,j), s_{i-1,j} + plt \right) \tag{2.1}$$

wherein

$$\delta(i,j) := \begin{cases} +1 & \text{, if } \varepsilon_1(i) = \varepsilon_2(j) \\ -1 & \text{, otherwise} \end{cases} \tag{2.2}$$

Here, $\varepsilon_k(b)$ denotes the b-th element of sequence k, and plt is a constant that weighs the costs for inserting a space into one of the sequences (typically, $plt = -2$, the double "price" of a mismatch).

The data dependences imposed by definition (2.1) imply a particular order of computation of the matrix: elements which can be computed independently of each other, i.e., in parallel, are located on a so-called *wavefront* which "moves" across the matrix as computations proceed. The wavefront is degenerated into a straight line when it is drawn along the single independent elements, but its "wavy" structure becomes apparent when it spans multi-element blocks. In higher-dimensional cases (3 or more input sequences), the wavefront becomes a hyperplane [bLa74].

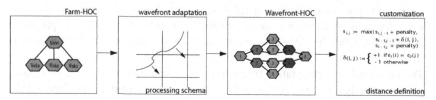

Figure 2.29 Two-Step Process: Adaptation and Customization

The wavefront pattern of parallel computation is used also in other popular applications: searching in graphs represented via their adjacency matrices, equation system solvers, character stream conversion problems, motion planning algorithms in robotics etc. Therefore, programmers would benefit from a Wavefront-HOC which captures the wavefront pattern. In this section, such a Wavefront-HOC is built by adapting the Farm-HOC and then customized to the sequence alignment application. Figure 2.29 schematically shows this two-step procedure. First, the wavefront adaptation is applied (which fixes a new processing order that is optimal with respect to the degree of parallelism, as explained below). Then, the distance definitions (2.1) and (2.2) are employed for customizing the adapted Farm-HOC for the sequence alignment application.

Creating the Wavefront-HOC
by Adapting the Farm-HOC

For the parallel processing of submatrices, the new Wavefront-HOC must, initially, establish the "wavefront order" for processing individual tasks, which is done by sorting the partitions of an initially empty matrix (the *workspace*) arranged by the Master, such that independent submatrices are grouped in one wavefront. In the presented case study, this sorted partitioning is computed while iterating over the matrix-antidiagonals as a preliminary step of the adapted farm, similar to the loop-skewing algorithm described in [bWo86] (Chapter 5 shows how the sorted iteration sequence looks in Fig. 5.11 and explains how the parallelzation can be automated). Only a single thread is uploaded to the original Farm-HOC as the adaptation code parameter for defining the wavefront processing schema. The wavefront-sorting procedure runs in the initialization method of this thread. After the initialization is finished, the adaptation code parameter keeps running concurrently to the original Master-thread and periodically creates new tasks by executing the loop shown in Fig. 2.30 in its run-method. This loop iterates over all wavefronts, i.e., the subma-

```
 1: for (List<Task> waveFront : data)  {
 2:   if (waveFront.size( ) < localLimit)
 3:     scheduler.dispatch(wave, true);
 4:   else  {
 5:     remoteTasks = waveFront.size( ) / 2;
 6:     if ((surplus = remoteTasks % machines) != 0)
 7:       remoteTasks -= surplus;
 8:     localTasks = waveFront.size( ) - remoteTasks;
 9:     scheduler.dispatch(
10:       waveFront.subList(0, remoteTasks), false);
11:     scheduler.dispatch(
12:       waveFront.subList(remoteTasks,
13:       remoteTasks + localTasks), true);  }
14:   scheduler.assignAll( );  }
```

Figure 2.30 The Adaptation Code Parameter for Wavefront Processing

trices positioned along the anti-diagonals of the similarity matrix being computed. The assignAll and the dispatch methods are not part of the standard Java API, but they were rather implemented specifically for the efficient scheduling of wavefront applications and work as follows: The assignAll-method waits until the tasks to be processed have been assigned to workers. Method dispatch, in its first parameter, expects a list of new tasks to be processed. Via the second boolean parameter, the method allows the caller to decide whether these tasks should be processed locally by the master (see lines 2–3 of Fig. 2.30): the adaptation code parameter checks if the number of tasks is less than a limit set by the client. All tasks of such a "small" wavefront are marked for local processing, thus, avoiding that communication costs exceed the time savings gained by employing remote servers. For wavefront sizes above the given limit, the balance of tasks for local and remote processing is computed in lines 5–8: half of the submatrices are processed locally and the remaining submatrices are evenly distributed among the remote servers. If there is no even distribution, the surplus matrices are assigned for local processing. Then, all submatrices are dispatched, either for local or remote processing (lines 9—13) and the assignAll-method is called (line 14). The submatrices are processed asynchronously, as assignAll only waits until all tasks have been *assigned*, not until they are finished.

Without the assignAll- and dispatch-methods, the adaptation parameter can implement the same behavior using a Condition from the standard concurrency API that was introduced with Java 5 for thread coordination. However, using only the standard Java API is a more low-level solution as compared to the abstraction level of the methods assignAll- and dispatch from the HOC Service API.

Customizing the Wavefront-HOC for Sequence Alignment

The HOC Service API (introduced in Section 2.3) includes several helper classes that simplify the processing of matrices. It is therefore, e.g., not necessary for the programmer to write the Master code for splitting matrices into equally-sized submatrices manually, but the API provides a standard procedure for a regular matrix splitting. The only customizing code parameter that must be written anew for computing the similarity matrix in the sequence alignment application is the Worker code. In this case study, this code parameter implements, instead of the general Worker-interface (Section 2.1.2), the alternative Binder abstract class (from the HOC APIs), which describes, specifically for matrix applications, how a matrix is initialized and how an element is computed depending on its indices:

```
1:  public abstract class Binder<E> implements Worker<E> {
2:    void init( );
3:    public E bind(int i, int j);
4:    public E[] compute(E[] input)  {  }
  }
```

Figure 2.31 Binder Interface for Matrix Worker Code

The compute-method implementation in line 4 is empty, since it is never called. Its only purpose is to enable compatibility with the Worker-interface, as explained in Section 2.1.4. Thus, the Binder abstract class does not provide its own functionality, but it is another version of the Worker-interface, which is more specific concerning the data being processes (indexed matrix elements, instead of arbitrary array entries).

Before the Wavefront-HOC computes the matrix elements, it assigns the empty workspace matrix to the code parameter; i.e., a matrix reference is passed to the parameter object and, thus, made available to the customizing parameter code for accessing the matrix elements. The customizing code parameter for calculating matrix elements (according to definition (2.1) from Section 3.4.2) initializes the border row and column of the workspace matrix with multiples of the gap penalty and its `bind`-method reads as in Fig. 2.32.

```
1:   public Integer bind(int i, int j)  {
2:   return max( matrix.get(i, j - 1) + penalty,
3:     matrix.get(i - 1, j - 1) + delta(i, j),
4:     matrix.get(i - 1, j) + penalty );  }
```

Figure 2.32 Customizing Code Parameter for Sequence Alignment

The helper method `delta` (line 3) implements definition (2.2).

The special `Matrix`-type used by the above code for representing the distributed matrix is also provided by the HOC API to facilitate location transparency, i.e., it allows to use the same interface for accessing remote elements and local elements. `Matrix` is an abstract class, and the HOC Service API includes two concrete implementations: `LocalMatrix` and `RemoteMatrix`. These classes allow to access elements in adjacent submatrices (using positive and negative indices), which further simplifies the programming of distributed matrix algorithms. Obviously, these API-specific utilities are helpful in the presented case study, but they are not necessary for adaptable components and therefore beyond the scope of our consideration.

Experiments with the Wavefront-HOC

The run times of the Wavefront-HOC and the original Farm-HOC were investigated using heterogeneous multiprocessor servers (located at a distance of approx. 500 km from the client) in order to study a realistic, grid-like scenario:

Server	Architecture	Processors	Clock Speed (MHz)
SMP U280	Sparc II	2	750
SMP U450	Sparc II	4	900
SMP U880	Sparc II	8	900
SMP U68K	UltraSparc III+	2	900
SMP SF12K	UltraSparc III+	8	1200

Table 2.2 The Servers in the grid Testbed

The input files used for the experiments were holding the genome data of various fungi, as archived at `http://www.ncbi.nlm.nih.gov`. The scalability was measured in two dimensions: (1) with increasing number of processors in a single server, and (2) with increasing number of servers. The diagrams in Fig. 2.33 and Fig. 2.34 show the results for computing the similarity matrix for two sequences of 1 MB size.

All diagrams show the average results of three measurements. To obtain a measure for the spread, the variation coefficient was computed for all test series; this turned out to be less than 5% for all test series.

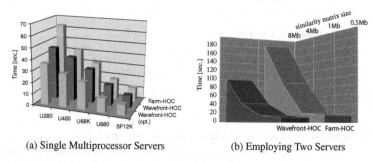

(a) Single Multiprocessor Servers   (b) Employing Two Servers

Figure 2.33 First Series of Experiments with the Wavefront-HOC

A standard, non-adapted farm can carry out computations on a single pair of DNA sequences only sequentially, due to the wavefront-structured data dependences. This behavior was imitated using the original Farm-HOC for the computation (without the adaptation parameter) and by specifying a partitioning grain equal to the size of an overall similarity matrix. This version was the slowest in all tests. Runtime measurements with the localLimit (line 2 of Fig. 2.30) in the adaptation code parameter set to a value $>= 0$ are labeled as *Wavefront-HOC (opt.)* in Fig. 2.33 (left).

The locality optimization explained above has an extra impact on the measured times since it avoids the use of sockets for local communication. To compare the Farm-HOC with the Wavefront-HOC without any optimization, the localLimit was set to zero in a second series of measurements, which labeled as Wavefront-HOC in Fig. 2.33. To investigate the scalability, the same application was run using two Pentium III servers under Linux. While the standard farm can only use one of the servers at a time, the adapted farm sends a part of the load to the second server which improves the overall performance when the input sequence length increases (Fig. 2.33, right).

For more than two servers the performance was degraded. This is probably due to the increase of communication for distributing the Binder-tasks (shown in Section 2.4.2) over the network. The diagrams in Fig. 2.34 support this assumption. The scalability was investigated using the U880 plus another SunFire 6800 with 24 1350 MHz UltraSPARC-IV processors. As can be seen, the performance of the application is significantly increased for the 32 processor configuration, since the SMP-machine-interconnection does not require the transmission of all tasks over the network. Curves for the standard farm are not shown in these diagrams, since they lie far above the shown curves and coincide for 8 and 32 processors, which proves again that this version does not allow for parallelism within the processing of a single sequence pair. The fourth diagram (Fig. 2.34, right) shows the effect of another interesting optimization: When the submatrices are compressed using the

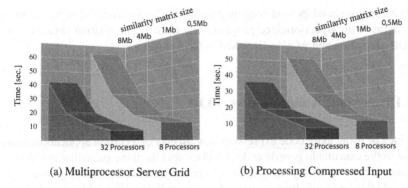

(a) Multiprocessor Server Grid   (b) Processing Compressed Input

Figure 2.34 Second Series of Experiments with the Wavefront-HOC

Java util.zip Deflater-class before they are transmitted over the network, the scaling behavior is very similar (the curves for the compressed data only grow a little bit slower for small-sized input), but the absolute times for larger matrices are considerably improved by the compressed submission.

Related Work

The use of the wavefront schema for parallel sequence alignment has been analyzed before in the CO_2P_3S system [bD⁺02], where it is classified as a design pattern. While in the CO_2P_3S system the wavefront behavior is a fixed part of the pattern implementation, in the HOC adaptation approach, it is only one of many possible adaptations that can be applied to a HOC.

The Wavefront-HOC was used for solving the DNA sequence pair alignment problem. In comparison with the extensive previous work on this time-consuming application (especially for large sequences [bK⁺02, bS⁺04]), the Wavefront-HOC represents a more user-friendly, high-level solution with competitive performance.

Due to the use of Web services for the communication, HOCs can be easily combined with other grid programming technologies. HOCs aim at abstracting over platform-specific issues in client applications, but an efficient server-side implementation of any grid component requires full control over the employed resources. Therefore, other grid programming technologies (including communication libraries, scheduling and job submission systems) are not an alternative to using HOCs. They rather represent complementary mechanisms which are of great value for grid system experts when new HOCs are designed. In the following chapters, some more complex HOCs than the Farm-HOC and the Wavefront-HOC will be presented and make obvious how other technical utilities for grid programming can be usefully combined with the HOCs and the HOC Service APIs shown in this chapter.

The component adaptation concept employed in the Wavefront-HOC has many similarities with aspect-oriented programming (AOP). The relation between component adaptation and AOP is discussed in the following.

2.5 Discussion: Adaptation vs. AOP

The idea of using extra code expressing new aspects of a certain application, in addition to the executable generic code that does not take these particular aspects into account, is not completely new. *Aspect-oriented programming* (AOP) tools like AspectJ [aLa03] (available from `eclipse.org`) or IBM's Hyper/J [cIB07b] address a similar issue: they extend object-oriented programming language by constructs for static and dynamic *crosscutting*. Crosscutting allows to capture concerns like logging and database transaction demarcation into *aspects* affecting multiple code blocks that usually spread across the established different divisions of responsibility in a software design. Imagine, e.g., a multi-tier architecture, where all the presentation issues are handled by a top-level tier of their own, but all GUI-events must be logged, probably in the same manner as the actions triggered in a lower-level application tier. While static crosscutting inserts additional methods or attributes into class definitions, dynamic crosscutting inserts into code so-called *advices*, which are additional actions taken whenever statements with certain properties are executed (e.g., a method invocation involving a given group of parameters).

The specific problem of component adaptation in the grid-context is that it deals with aspects of software that is deployed into a remote hosting environment, after this environment has been launched.

Besides the capability of applying adaptations at runtime, the use of code parameters in HOC adaptation also has the advantage that this technique does not cause a dependence between adaptations and a particular AOP tool. The syntactic representation of aspects is not standardized, but rather depends on the employed tool. The HOC model couples adaptivity only to the service interface via an additional code parameter. Thus, the HOCs and the code parameters for adapting them neither depend on the use of a tool like AspectJ nor on a particular programming language, whereas AOP is restricted to object-oriented technologies only.

Aspects and adaptation can be viewed as complementing each other. For modifying the treatment of finished tasks in the Wavefront-HOC, the component takes advantage of the fact that the master of the original Farm-HOC publishes state changes, such as task terminations, according to the *observer pattern* [aG⁺95]. When a farm component other than the Farm-HOC, which does not adhere to the observer pattern, should be adapted, the typical observer behavior can be added via dynamic crosscutting. Static crosscutting enables an elegant way for extending any farm implementation by methods, such as `assignAll` and `dispatch`, allowing additional threads to interact with the farm master.

Of course, HOC adaptation is not restricted neither to the Farm-HOC nor to wavefront algorithms. In an adaptation of other HOCs like the Divide-and-Conquer-

HOC from the HOC-SA repository (shown in Chapter 3), the technique can be used analogously: If, e.g., an application of a divide-and-conquer algorithm allowed to conduct the join-phase in advance to the final data partitioning under certain circumstances, this optimization can be applied using an adaptation without any impact on the standard division-predicate of the algorithm. Adaptations work by supplying a HOC with only a part of the code parameters it accepts, producing a new HOC with an alternative behavior. More research on higher-order constructs and, so-called *partial functional applications* is conducted in the skeleton research community [bKS02, bPK08].

Chapter 3
Higher-Order Component Service Architecture (HOC-SA)

Web services are the communication technology which is used for connecting HOCs and client applications over the Internet in all examples presented in this book. In Chapter 2, HOCs were introduced and the use of Web services was taken for granted, since Web services are the standard communication technology in the Globus toolkit which is the underlying grid middleware of all current HOC implementations. In this chapter the advantages of Web services over alternative communication technologies, especially concerning the interoperability among software which is written in different programming languages, are discussed in more detail.

Interoperating components, implemented in multiple programming languages, is one of the key features of service-oriented architectures (SOAs [aEr04]). Web services facilitate the implementation of such architectures, since the formats of all data which can be used as valid input or output of a Web service are clearly defined. On the other hand, the most obvious shortcoming of Web services is the costly encoding and decoding of XML documents required for every data transmission, which is a serious problem for performance-critical grid applications. This book suggests HOCs which combine Web services with other communication technologies as a solution to this problem.

As an example for a HOC which integrates multiple communication technologies the so-called *Lifting-HOC* (the origin of this name is its implementation structure explained in Section 3.4.2) is developed in this chapter. Client applications can use the Lifting-HOC for applying filters (based on the discrete wavelet transform) to input data (number series or images) and run the time-consuming filtering process on the grid. A Java-based Web application which is connected to the Lifting-HOC via a Web service is used for uploading input data from the client [bD+05]. The server-side processes which perform the actual data processing make use of the C/MPI-based skeleton library *eSkel* [cCB07].

The outline of this Chapter is as follows: Section 3.1 illustrates the HOC-SA programming schema. Section 3.2 clarifies the relation between services, resources and components. A detailed comparison between the HOC-SA and the Globus Resource Allocation Manager (WS-GRAM [bC+98]) which is the standard Web service for code transfer in the Globus Toolkit, follows in Section 3.3. Section 3.4 presents

J. Dünnweber, S. Gorlatch, *Higher-Order Components for Grid Programming*, DOI 10.1007/978-3-642-00841-2_3, © Springer-Verlag Berlin Heidelberg 2009 a solution for integrating MPI-based skeletons (and, potentially, other non-object-oriented, self-contained software) with HOCs using a *gateway* – a program with the only purpose of accessing an external system (MPI, in the example) from within a component. A new case study application is introduced: the discrete wavelet transform (DWT) as it is used, e.g., in image processing [bC$^+$06]. The case study shows how customizations and adaptations can be applied to HOCs written in programming languages other than Java. A performance evaluation of the gateway in the DWT application is also conducted. Section 3.5 shows how multiple HOCs can be combined in a single HOC-SA application that adheres to the Map-Reduce pattern of parallelism Section 3.6 summarizes the most important features of the HOC-SA.

This section explains how to work with the Higher-Order Component Service Architecture (HOC-SA), our implementation of a Service-Oriented Architecture (SOA) that enables users to run grid applications based on Higher-Order Components (HOCs) on top of the latest Globus middleware. The HOC-SA comprises a component repository, which offers to the client several HOC implementations plus the necessary means for accessing them remotely via Web services. These means are the Code Service and the Remote Code Loader. The Code Service is a specially configured Web service for maintaining code units in grid-wide accessible databases. The Remote Code Loader allows programs to load code units stored in the Code Service over the network and run them like local code.

One objective of this book is to offer a high-level programming model for the grid which shields the programmer from low-level details (interactions between multiple services and the parallelization strategies), thus allowing to concentrate on the application logic. HOCs help achieving this goal. In this chapter, the service-oriented architecture used as runtime environment for HOCs is described in more detail. For data sharing among HOCs, the OGSA-DAI framework (Data Access and Integration [cUD07])) is used, which facilitates a uniform interface to databases in the grid.

3.1 Service-Oriented Grid Programming Using the HOC-SA

The HOC-SA is a further development of the general approach known as Service-Oriented Architectures (SOA) [aEr04]. SOA systems are composed of distributed services, which are not necessarily Web services, but some network-accessible entities, capable of serving requests that arise recurrently in different applications [cEr06]. Another significant property of services (contrary to other software) is that they have well-defined interfaces and dependencies specified in standardized, usually XML-based, formats. Currently, the most common SOA implementation technologies are either plain Web services as defined by the W3C [cW302] or, in the grid context, Web services and so-called Web service resources as defined in the Web Service Resource Framework (WSRF [cOA04])). The role of Web service resources in the HOC-SA is explained below.

In a Service-Oriented Architecture (SOA), computing entities interact in a way that enables one entity to perform work on behalf of another entity [aEr04]. SOAs can be built using various distributed computing technologies like, e.g., CORBA, .NET or Web services. Since Web services allow for distributed computing in an Internet-wide setting, SOAs based on Web services offer clear advantages for the development of grid applications. Hence, emerging standards for grid computing make use of Web services or are actually designed to be incorporated into Web services standards like the Web service Resource Framework (WSRF). The steady integration of such new standards into grid computing creates new potentials, but at the same time poses challenges for application programmers who have to cope with an increasing number of technologies. HOCs enable programmers to benefit from many of contemporary grid technologies without having to deal with all of them directly.

The main steps of running a HOC-based program which uses the HOC-SA for transferring code are depicted in Fig. 3.1:

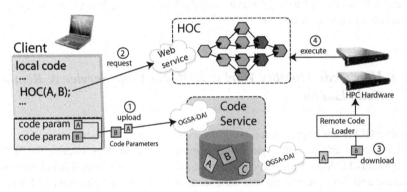

Figure 3.1 Code Transfer to a Component (HOC) via the HOC-SA

step ①: In this upload step, the client stores code parameters, i.e., the application-specific part of the program, in the Code Service (a Web service connected to a database via OGSA-DAI [cUD07], see Section 3.1.1). The result of this step is that the code parameters ($\boxed{A}$ and $\boxed{B}$ in the figure) are saved together with unique identifiers (A and B).

step ②: When execution a HOC-based program, the client invokes a HOC by performing the call HOC(A,B) and provides the code parameters for the application by sending the identifiers A and B. Principally, this is an ordinary Web service request but it is served by a HOC: the client connects to only one Web service, but serving the request may involve multiple remote components. For each HOC, there is one Web service making it publicly available. Thus, connecting, e.g., to the Farm-HOC Web service implies that the request is served by the Farm-HOC and the code parameters $\boxed{A}$ and $\boxed{B}$ are executed within this context (i.e., as Farm-HOC Master and Worker).

step ③: In the download, the code that the identifiers from the invocation in step ② refer to is transferred to the HPC hardware (using OGSA-DAI for the retrieval) and linked to the HOC using the HOC-SA Remote Code Loader (explained in Section 3.1.1).

step ④: The instantiated HOC is executed in a parallel, high-performance manner using the selected target machines. This selection is either done in the client code by the user, who directly calls `configuregrid(serverList, ...)`, as shown in Chapter 2, or implicitly by an advanced scheduler, as explained in Chapter 5.

Web service requests to a HOC (such as ② in Fig. 3.1) are served asynchronously with the client execution in the HOC-SA. The required setup for the asynchronous Web service call (using notifications, as defined in WS-N [cOA06]) is provided by the HOC-SA, allowing client programmers to benefit from asynchronous processing without dealing with notifications, in the client code.

The number of grid servers involved in the execution of a HOC is not limited. Actually, it can also be a single server, and the Code Service is not even necessarily hosted on a machine of its own.

3.1.1 How Code Mobility Works: HOC-SA Code Service & Remote Code Loader

Because HOCs are implemented on grid servers while the application-specific customizing code units (HOCs' parameters) must be provided by clients, the HOC-SA contains facilities for *code mobility*, i.e., shipping code units from clients to a database and from there to servers for execution. A high-level mechanism for code mobility was proposed in Section 3.4.9 and explained in a programming language-independent manner. Here, we explain how code mobility works in our Java-based implementation of the HOC-SA.

Web services communicate using plain SOAP. Thus, it is not possible to pass an argument of a type that cannot be declared using an XML-Schema [cW396]. Therefore, Web service parameters may be primitives or data records declared in a class-like manner, including associations and inheritance, but code in the form of method implementations cannot be transferred using traditional mechanisms. For the Farm-HOC, the solution to this problem using a `Master` and a `Worker` reference, instead of passing directly the code to the component has already been shown in Section 2.1.3.

Code mobility in the HOC-SA is implemented by representing the code as primitive data (arrays of bytes which can easily be transferred via SOAP) and passing references to this data to any HOC. Besides the possibility to declare code as any other data in the Web service interface, its representation in an array makes it possible to use a special transfer mechanism for contiguous data (WS Attachments [aEr04]) which is more efficient than XML serialization. For maintaining code units persistently and reusing them in different applications, an advanced storage and retrieval mechanism is needed.

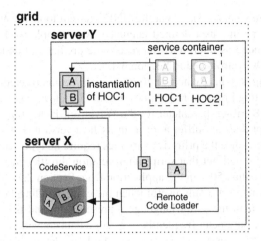

Figure 3.2 Code Service and the Remote Code Loader

Therefore, the HOC-SA includes the Code Service which uses Web services for exchanging code over the network and a database management system as backend. Figure 3.2 shows how the Code Service and the Remote Code Loader work together in an application using two HOCs. Server Y runs a service container (typically Globus) hosting the HOC repository which contains the HOCs HOC1 and HOC2 (for the sake of simplicity, each of these two components consists of a single service that is local to server Y). On server Y, the standard Java class loader is replaced by the Remote Code Loader that downloads from the Code Service those class definitions which are not available on the local file system (the code units $\boxed{\text{A}}$ and $\boxed{\text{B}}$ in the figure), and creates instances of the classes defined therein using the Java reflection mechanism [cSM07c]. This way, the transferred codes, that have been encoded as byte arrays during the transmission, become Java objects again. When server Y then processes a service request by performing the operation specified in the invocation, these objects perform the application-specific work.

While other code transfer mechanisms exist (e.g., Job submission systems like the Sun grid Engine [aC+04]), the HOC-SA has the advantage that it supports the reuse principle of component-based software design: once a class has been loaded, applications can create as many instances of it as needed without transferring code repeatedly. In the rest of this chapter, some additional advantages will become obvious, especially in Section 3.3, where HOC-SA is compared to WS-GRAM, the standard code transfer technology in the Globus toolkit.

The HOC-SA code transfer mechanism supports five different formats of mobile code which can be used by the application programmer:

1. **Java bytecode** which is the most common format, expected by HOCs as the default encoding of code parameters. This is also the format which is used in all the examples in this book, where the use of a different format is not explicitly mentioned.
2. **Interpreted terms**, which can be evaluated using different tools, e.g., the JEP Parser [cSi06] which can interpret arithmetic terms (as they are used by the

Lifting-HOC in Chapter 3), or via the UNIX system tools *lex* & *yacc* [aMB90] for parsing terms in a user-defined language. An example for HOC parameters which are encoded as interpreted terms are the predict and update functions in the Lifting-HOC, introduced in Section 3.4.5.

3. **Script languages,** e.g., Python or Ruby, which are interpreted remotely using the Bean Scripting Framework [cAp04]. Figure 3.3 shows an example: the Ruby version of the `Binder`-parameter for the Wavefront-HOC from Section 2.4.

    The short script does not differ a lot from its Java version (Fig. 2.32), but it is a little bit simpler, since the primitive variables require no type declarations. It was experimentally found out that script interpretation throttles the application performance (more than 50% in the application from Section 2.4). Therefore, script language parameters are only useful when the corresponding Java code is much more complex than the script language representation and the employed HOC does not run the parameter frequently (e.g., for a data conversion or initialization step, but not for processing every element of a large data structure as the matrix in the example).

```
def bind(i, j)
   return [ $matrix.get(i, j-1) + $penalty,
            $matrix.get(i-1, j-1) + p(i, j)$,
            $matrix.get(i-1,j) + $penalty ].max
end
```

Figure 3.3 Ruby Version of the Code Parameter for Sequence Alignment

4. **Native binary code** for a platform specified by the user. As a consequence of using this non-portable format, only one particular execution platform can be used. This format is transferred using a code cut-out technique as shown for an example in Section 3.4.9.

5. **Source code** which can be written in Java or C/C++. Using this format implies that the compiler and all libraries required for translating the sources are available on the execution host (none of the examples shown in this book uses this format).

The format of each code parameter which is not encoded using the default Java bytecode format must be specified by the HOC-SA user when the parameter is uploaded to the Code Service. This specification is done in step ① of Fig. 3.1 and clarifies where and how a code parameter can be executed, whereby the code actually becomes *mobile* in the sense of this book (contrary to a *portable* code, which is supposed to be executable anywhere). Another possibility for transferring code in a non-portable format would be to compile the code for a particular host's processor type on the client, using a cross-compiler, and to store the machine code to the Code Service, but this possibility is not implemented by the HOC-SA.

The databases used for storing the code parameters save the format of each code parameter as a `type`-attribute together with the code (e.g., in an extra column of the same table row, when a relational database is used). These databases and their connection to the HOC-SA are now described in more detail.

3.1.2 Parameter Databases in the HOC-SA

Code parameters provided to a HOC by uploading them to the Code Service can be reused in multiple applications, since they are stored persistently.

Persistence and Code Reuse

An example where the reuse of code parameters is obviously useful is an image processing application that applies the same filter to multiple images: this application can be built by nesting Farm-HOCs into a farm of farms, such that each single image is processed using a farm of multiple processors, while all images are processed by such farms in parallel. Instead of transferring the sequential filtering code for each particular farm repeatedly, the client rather uploads this code only once to the Code Service, making it available to all workers (Chapter 5 explains in more detail the relation between HOCs and the CoreGRID GCM [cCN05] model which is ideal for programming applications with nested components due to its hierarchical composition features).

Another example are applications that use partially parameterized HOCs, i.e., HOCs that have already been provided with a portion of the required code parameters in advance. Imagine different image processing applications: one application may produce fractal images; another application may apply filters to these images and thereby add effects like blurring, fading or glowing. Both applications can be implemented using the Farm-HOC with the same master code loaded from persistent storage. Once a programmer has supplied the Code Service with the master code unit for image processing, the Farm-HOC can be used as a generic image processing service customizable for a particular application by means of the worker code. For recurring applications like image processing, also the worker unit does not have to be written anew each time when a standard filter is needed. Instead it can be stored persistently and reselected, e.g., once for the Farm-HOC from Chapter 2 and another time for a stage of the Lifting-HOC from Chapter 3, as a part of a multi-stage filtering operation. A more extensive examination of programming with partially parameterized higher-order constructs and combinations of them has been conducted in [bKS02].

Use of Multiple Databases

The HOC-SA exploits the OGSA-DAI framework to store code units persistently in databases, since OGSA-DAI provides a standardized means for connecting grid applications to databases via Web services. Using OGSA-DAI, grid applications can query, update, transform and transfer database contents.

OGSA-DAI provides grid-wide transparent access to (possibly distributed) databases, i.e., the Code Service can access a code unit without caring about the actual location and type of the underlying database system using the standard DAI data retrieval mechanism.

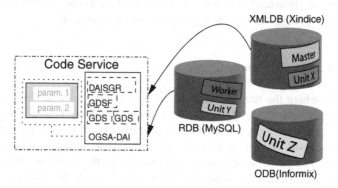

Figure 3.4 Retrieving Code Parameters from Multiple Databases

Different kinds of databases can be useful in the HOC-SA. This is shown in Fig. 3.4 (DAISGR, GDSF and GDS are parts of OGSA-DAI, explained below). If a programmer wants to search and reload code units without knowing their names, providing only semantic information, querying an XML-database like Xindice might be preferred, which allows to classify data using an ontology. Ontologies can be described, using, e.g., Topic Maps [aPH04] or the Resource Description Framework (RDF [cW304]) and allow users to organize data by associating it with metadata [bB⁺01]. Another useful type of databases for mobile code units are object-relational systems like Informix: they can reload associated code units (e.g., the `predict` and `update` stage for the Lifting-HOC in Chapter 3) as aggregated entities, thus, reducing remote communication. A plain RDB like MySQL that stores code units as primitive CLOBs (**C**haracter **L**arge **Ob**jects) can also be used to store code units. For security reasons, databases are specially configured to ensure that only designated clients are allowed to store or edit code units. Thus, HOC developers have not to be concerned with security issues (which, of course, are an important aspect when exchanging executable code): this job can be delegated to database administrators.

The OGSA-DAI Retrieval Mechanism

The HOC-SA Code Service encapsulates a query evaluation mechanism, such that a programmer using the HOC-SA does not have to deal with OGSA-DAI directly. However, OGSA-DAI is a powerful technology that allows for queries to relational databases with simple keys as well as for XML-databases supporting query languages like XPath [cW399] and XQuery [cW307]. To make these features available to the programmer, the Code Service allows the HOC-SA user to interact with the

OGSA-DAI services (see Section 2.3 for code examples showing how this is done using the HOC Service API).

The central access point for all databases in OGSA-DAI [cUD07] is a GDS (Grid Data Service) that is used to submit queries to any kind of database. To make the programmer independent from concrete database types, the architecture of GDS follows the abstract factory design pattern [aG+95] as follows: a GDS-factory can be configured for a specific database, but it always creates GDS instances with a common interface. In OGSA-DAI, this common pattern is extended by a mechanism for including databases in the grid dynamically: new databases are registered using a service called DAISGR (DAI Service Group Registry). Using the OGSA-DQP (Distributed Query Processor), the Code Service allows the user to run distributed queries.

3.2 HOCs and Web Services

Web services and components are both programming technologies which subdivide a distributed system more precisely than just into a client and a server program. Both technologies have different features, especially with respect to the network communication. Most traditional middleware systems support either Web services or components which use a proprietary communication protocol. The HOC-SA is an extension to the Globus middleware, allowing its users to combine both: (1) components (HOCs) interconnected, e.g., via TCP-sockets or RMI, and (2) Web services making the HOCs available to clients via the Internet.

A closer look at both technologies makes obvious that this combination is a good compromise between the maximum possible interoperability and the minimum necessary costs for data exchange.

3.2.1 Web Services

Web services, as defined by the W3C standardization consortium, are "designed to support interoperable machine-to-machine interaction over a network" [cW302]. Technically, this interoperability is realized by decoupling the services' functionality from the underlying support system; this allows to access services in an implementation-independent manner. A Web service is a server application that can be requested to perform operations on behalf of a *consumer* that is either a client program or another Web service. Contrary to traditional server applications, Web services are not self-contained programs. A feature, which is often mentioned as an advantage of Web services over self-contained server programs, is that they are more light-weight, since the communication code is not a part of the service code but handled by a hosting environment that connects services and consumers. Such hosting environments that can host several different services are Web service containers (a special type of the container software, introduced in Chapter 1); Web service containers fall into the category of middleware forming a layer between operating systems and applications.

According to Szyperski's definition [aSz98], Web services can be viewed as a kind of software components. However, most HOCs in the HOC-Service Architecture are components which are much more complex than a single Web service: they are composed of multiple interconnected Web services or other components, while the Web services are just used for communicating between HOCs and clients or among multiple HOCs.

Any container software extends the functionality of the components deployed therein (i.e., installed with a proper configuration) by application-independent features. Such extended functionality, which can be useful in the context of grid computing, may include, e.g., failover procedures or the encryption and decryption of sensible data etc. The most basic functionality which a Web service container must provide is the transparent marshaling and unmarshaling of parameters and return values communicated over the network. Requests to remote services in the client code and the server-sided code for processing them look like the code for local method invocations (or like a subroutine call in a non-object-oriented programming language). Behind the scenes, the container converts all data transmissions into the XML-based SOAP format, allowing to interconnect distributed software in heterogeneous networks.

Web service containers handle every request for Web service operations by spawning a new thread (or taking one from a thread pool). Thus, a Web service container allows multiple client programs (or grid components) having their requests served simultaneously. Programmers who implement the service operations do not need to deal with coordinating the threads, since the container schedules them. Web service containers handle the transmission of SOAP-encoded data (request parameters and results) using HTTP as the wire protocol via the standard port 8080 (if not configured differently by the user). Such transmissions pass across most firewalls and Internet proxies, which is obviously another advantage which such a container provides to Web service-based grid applications.

Many different middleware systems which include (or consist in) a Web service container exist, e.g., commercial application server programs (such as *WebLogic* [cBl06] and *WebSphere* [cIB03a]) and the *Apache Axis* open-source system [cAp03b]. The Globus Toolkit (in its most recent versions, i.e., version 4 or higher) which is used as the underlying middleware of the HOC-SA, provides an open-source Web service container which is called Globus container and complies to the WSRF standard [cOA04]. Since all HOCs maintain some data using WS resources (see, e.g., Section 2.1.4 for an example) defined by the WSRF standard, any of the commercial Web service containers (which, typically, do not use WS resources) would require a more complicated setup than Globus. Thus, all our experiments were conducted using the Globus container. WS resources are important for combining Web services with other software components, as explained in the following.

3.2.2 Components and Resources

While Web services themselves are a special kind of components (as explained above), there is a difference between Web services and most other software components: Web services depend neither on the context of the calling application nor on its state. Consequently, there is no particular order in which the operations offered by a service must be requested. This feature allows to replace a Web service with another one in-between two operation requests, without affecting the consumer software (e.g., the client). This is an advantageous property in grid applications, where it helps addressing possible failures of the distributed hosts.

However, Web services also have two serious drawbacks as compared to the other (typically more complex) software components (HOCs or others) in distributed systems: (1) a Web service is not only state-independent, but it is fully *stateless*, i.e., it cannot maintain any data values outlasting a single operation, and (2) a Web service is *intransient*, i.e., once deployed to the container, it is initialized when the container is launched, and not shut down before the container stops. Therefore, consumers cannot create multiple independent instances of a Web service and distinguish them. Once multiple consumers access the same data resource (e.g., a database) via a Web service, simultaneous requests will quickly cause conflicts, since the service cannot distinguish between the consumers.

To overcome these drawbacks, the Open Grid Service Infrastructure (OGSI) [cTT04], which is outdated now, defined the notion of *grid services*, a special type of Web services that are *stateful* and *transient*. Grid services could be instantiated and destroyed on demand. Unfortunately, OGSI replaced common Web service standards like WSDL and WSDD with OGSI-specific derivatives. These non-standard solutions (GWSDL and GWSDD) hampered the acceptance of Globus Toolkit 3 and were abandoned in the Open Grid Service Architecture (OGSA) provided by the recent Globus Toolkit versions [cGF04]. As obvious from the OGSA acronym, the term "grid services" is still used in the context of Globus, also in version 4 and higher: it now refers to a Web service that maintains application state data accordingly to the Web service resource framework (WSRF) specification [cOA04]. Such grid services are stateless and intransient like any other Web service.

The new Globus grid services are used within a *façade* [aG+95], which is a data management abstraction composed of the WSRF-typical Web service resources. WS resources represent software entities that have their own configuration (i.e., data attributes defined in XML documents during deployment) and can be accessed individually. Users pass to the façade so-called *endpoints*, which allow the Globus container to distinguish resources using WS Addressing [cAp06]. Any software that is exposed in the Internet via a Web service using such a WS resource represents a full-fledged software component, i.e., a component supporting the conflict-free remote processing of data, regardless of the number of involved clients and operations.

The term grid service is ambiguous, since in the OGSI implementation it was something completely different than in WSRF. Therefore, in this book, the term grid service is not used outside this section. Instead, also the new grid-aware Web services (using WS resources) are called Web services or simply "services," wherever the implementation technology does not matter.

3.2.3 The HOC-SA Component Repository

The HOC-SA Component Repository [bDG05] is a kind of remote library offering multiple HOCs for reuse in different grid applications. Four of the HOCs in this repository have already been introduced in the examples in the previous chapters (Farm-HOC, Pipeline-HOC and Wavefront-HOC in Chapter 2, and the Lifting-HOC in Chapter 3). To illustrate the idea of implementing general-purpose recurring patterns as HOCs a bit more, this section lists the remaining six HOCs currently contained in the repository. Two more, relatively simple HOCs are:

- The *Reduce-HOC* implementing the reduction pattern of computation, where an application-specific binary operation is used to combine multiple data into a single result value. The code unit implementing this binary operation (e.g., addition on numbers) is the code parameter of the Reduce-HOC.
- The *Divide&Conquer-HOC* offers the divide-and-conquer pattern and requires four code parameters: One for the divide phase, one for the combine phase, one for processing the base case and, finally, a predicate deciding when the base case is reached.

The current HOC-SA repository also holds components with a more complex, but also more specific functionality: the *Alignment-HOC* for bioinformatics and the *Deformation-HOC* for 3D-simulations (application examples for both follow in Chapter 4). Chapter 5 introduces the *LooPo-HOC*, a special version of the Farm-HOC with embedded loop parallelization.

The basic functionalities which are commonly provided by all HOCs in the HOC-SA component repository are:

1. a HOC can be parameterized with data and code from the Code Service;
2. a pool of grid servers can be specified and thereby made available to the HOC for outsourcing computations;
3. every HOC has one operation to initiate a new (asynchronous) process and one operation to retrieve results.

The Java interfaces for using this functionality were already shown, as a part of the HOC Service API in Section 2.3.

A very simple HOC is the Worker HOC. There is no parallel processing pattern included in the Worker-HOC. Its only purpose is to provide a basis allowing users to create new HOCs and extend them to the HOC-SA component repository by inheriting the three basic functionalities from it and implementing new processing patterns. Using the Farm-HOC in a similar manner for deriving new HOCs with a farm-like structure from it was already discussed in Chapter 2.

3.2.4 The HOC-SA Portal

For developing large-scale applications, a comfortable HOC-SA portal has been designed. This portal can be used for choosing HOCs, for uploading, compiling and managing code parameters, and for customizing and controlling applications online, using a standard Web browser.

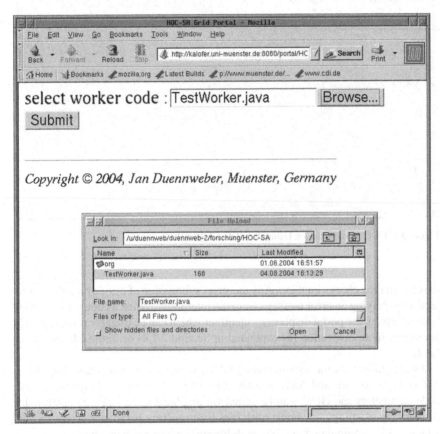

Figure 3.5 HOC/Parameter Selection in the HOC-SA Portal

Figure 3.5, Fig. 3.6, Fig. 3.7 and Fig. 3.8 show four screenshots of the HOC-SA portal.

Figure 3.5 (a) is one of the starting pages which are composed of plain Web forms (based on Jakarta Struts [cAp03c]). Once a HOC is chosen, the user can submit code parameters using an HTTP-upload, i.e., the browser opens a file selector that allows to select a local file and transmit it via HTTP. This user interface represents a distributed, interactive development environment for programming HOC applications. It allows programmers to browse, display and edit code parameters or

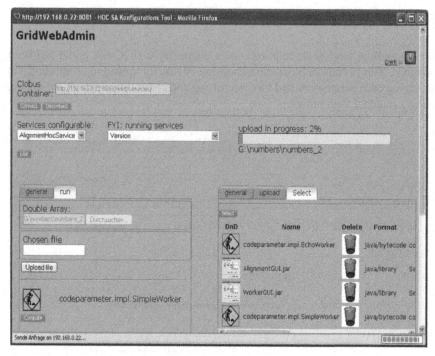

Figure 3.6 HOC Administration Using the HOC-SA Portal

to perform type checks on them. Any incorrect submission of a code parameter is re-
jected, displaying an error message from the interpreter. When a code parameter has
been accepted, the portal displays its identifier which can be used for referring to the
parameter in an invocation (as explained in Section 2.1.4) or in the administration
GUI.

Figure 3.6 shows the administration GUI which was implemented using Asynchronous JavaScript and XML (AJAX [cMD08]). Here, the assignment between
code parameters and HOCs can be edited via graphical icons (such as construction
workers representing code units, trash bins for triggering parameter deletions, etc.)
which can be manipulated via drag-and-drop in the browser window.

Figure 3.7 and Fig. 3.8 show the parts of the portal which offer the main possibilities for user interactions: two example GUI tabs for application configuration.
These tabs represent application-specific user interfaces (for a bioinformatics application, in the depicted examples) and are assigned to the HOCs in a similar manner
as code parameters.

Since the number of code and data parameters and their types differ in HOC-
based applications, a single customization GUI can hardly cover all possible applications. Therefore, a more flexible GUI design was conceived for the HOC-SA
portal, which makes use of multiple interchangeable GUI tabs for customizing and

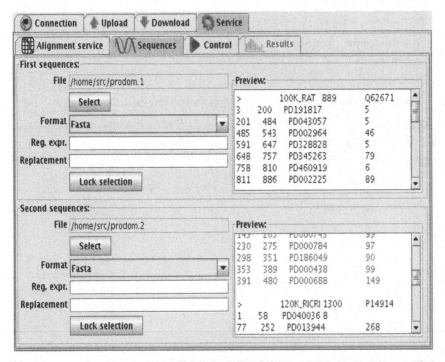

Figure 3.7 Input Selection for an Application in the HOC-SA Portal

controlling different HOC applications. An example that shows how users can assign GUI tabs to HOCs using the HOC-SA portal is shown in Fig. 3.6. Two `jar`-archives are shown in the box in the lower right corner of this screenshot of the administration GUI, the `AlignmentGUI.jar` and the `WorkerGUI.jar`. These `jar`-archives hold the Java code of GUI tabs and are depicted as thumbnail-size previews of the GUI tabs they implement: `AlignmentGUI.jar` contains the code of the GUI tabs for the genome sequence alignment application, shown in Fig. 3.7 and in Fig. 3.8. By placing the preview symbols next to the appropriate HOCs, administrators can graphically select the GUI tabs provided to the users for customizing the HOCs chosen for their applications. GUI tabs are Java Swing applications that can be launched from within the browser using Java Web Start [cSM05b].

All parts of the HOC-SA portal, the Web interface for selecting HOCs and code parameters, the administration GUI for editing assignments between HOCs, code parameters and GUI tabs and the applications-specific GUI tabs are implemented as independent Java Web applications. Thus, users can freely decide for each step in HOC-based application development and execution, if they handle this step using the graphical HOC-SA portal or directly in a client program (as, e.g., the Clayworks client, in Section 4.1.1).

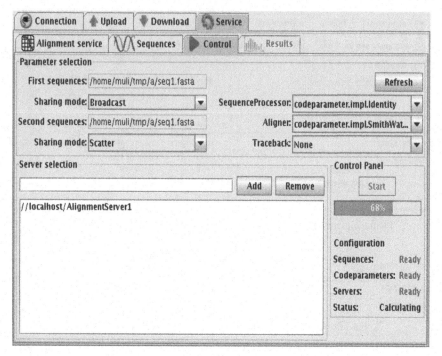

Figure 3.8 Application Progress Control in the HOC-SA Portal

3.3 A Comparison of the HOC-SA and Globus WS-GRAM

Transferring code over the network is fundamental to grid programs. Grid hosts often run activities which are controlled, or even defined, by other computers sending them code (even in systems that do not use any kind of software components).

The objective of the HOC-SA is to aid programmers with a programming model using components which simplify programming by abstracting over code and data transfer. The WS-GRAM tool in Globus also allows programmers to dispatch jobs to remote computers using Web services, but WS-GRAM still exposes to its users more details of the grid than the HOC-SA (e.g., with respect to the job declarations which WS-GRAM users must write in RSL format [bC+98]).

When the Globus middleware is used for performing the communication between heterogeneous hosts in an application, it requires that programmers write detailed declarations for every piece of code and data which is exchanged over the network. Thus, programming a grid application usually consists of writing an operational part and a declarative part. The operational part handles the actual computation work in the application. This part is quite complex, since the computations run in parallel on several network nodes which the programmer must coordinate. The declarative part has the purpose of instructing the middleware how to handle the different aspects of the network communication. This part has multiple elements including various resource specifications, interface definitions, deployment descriptors, etc., each in a middleware-specific format (see Chapter 2). The division of programming tasks originates from distributed business applications, where also the use of Web services for communication has been established [cIB03b] before it became a common middleware standard.

WS-GRAM [bC+98] is currently a de facto standard for transferring code, packaged as a job in the grid using Web services. In WS-GRAM, a job is a Web service parameter of a special type. Physically, a job is an executable program, packaged together with the required information to run this program: a description of the parameters this program expects and the program's requirements concerning the processors and libraries available on the execution platform. WS-GRAM extends the Web service standards (WSDL and SOAP [cW302]) by a descriptive language for job definitions, since usually the types used for the parameters of Web services are plain data types but not executable programs. The Resource Specification Language (RSL [bC+98]) is the XML-based format for describing jobs in WS-GRAM.

3.3.1 Grid Programming with WS-GRAM and the HOC-SA

To execute a program, the execution host requires, beside the binary code, some context information describing the libraries used by the program and the parameters it expects (including their type and order). If a program is transferred to a Web service, it is encoded like plain data, i.e., the context information is not automatically supplied to the Web service provider.

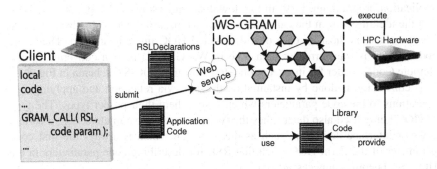

Figure 3.9 Code Transfer via Job Submission in WS-GRAM

Using WS-GRAM

Figure 3.9 shows how full programs are transferred in WS-GRAM: The client code (left part of the figure) contains a call to WS-GRAM where application code is submitted as a job to the execution hosts. The colored hexagons in the WS-GRAM job represent processes connected by arrows representing message exchange.

Besides the application code, the client sends the RSL declarations of the job. The programmer must write these declarations to provide the information about how to invoke the transferred code at the execution hosts. RSL files are uploaded by WS-GRAM users at runtime together with the code described therein. RSL belongs to the declarative part of a WS-GRAM application. Thus, RSL extends the static configuration which any Web service requires (WSDD & WSDL [cW302]) by a dynamic configuration part. While the introduction of a dynamic part to the Web service configuration enables the use of different kinds of executable code as parameters, it requires users to be familiar with all the service configuration. Additionally to the parameters for their applications, WS-GRAM users must describe the libraries used by their jobs (which must be provided by the execution hosts) in RSL-format. In Fig. 3.9, there is one RSL declaration document per application code unit (i.e., per unit transferred in a single submission, typically a class).

Using the HOC-SA

The transfer of mobile code in the HOC-SA was already explained in Section 3.1. The main difference to WS-GRAM is that the transferred codes do not represent full programs (directly executable on the remote host), but rather code parameters which are executed in the context of a HOC which has all the required context information [bD⁺07].

For transparently inserting code parameters into the appropriate positions in the pattern implemented by a HOC, the HOC-SA performs two steps invisible to the application programmer. First, in the download step (step ③ of Fig. 3.1), the code that the identifiers from the call refer to is transferred to the HPC hardware (which is more than one host, in case of a distributed HOC implementation). There, the code (which is transferred as data, like in WS-GRAM) is then converted by the Remote Code Loader for its execution (step ④ of the HOC-SA schema in Fig. 3.1). This conversion is done by instantiating the code via reflection and applying cast operations to the code parameters which assign them their proper types. The type of HOC being used also determines the type of the code parameters: e.g., a certain class or interface definition, such as the Master which is always the first code parameter of the Farm-HOC, such that RSL for describing code parameters in the HOC-SA becomes unnecessary.

3.3.2 Application Types for HOC-SA and WS-GRAM

Both WS-GRAM and HOC-SA delegate the handling of grid-specific require-ments to the Globus middleware [bFo06]. Applications may use any of the ad-vanced features in Globus, e.g., the Reliable File Transfer service (RFT [bM⁺02]) which provides a basis for fault-tolerance, and the Globus Security Architecture (GSI [bW⁺03]) which ensures that communication takes place only among trusted partners. Thus, the middleware offers the same variety of functions to WS-GRAM and HOC-SA users.

In the following examples, best practices for code transfer are demonstrated, either using HOCs that simplify the use of the middleware or via WS-GRAM which gives programmers more control over the middleware and may lead to better performance (a detailed performance comparison follows in Section 3.3.3).

Astronomic Sky Survey with WS-GRAM

The AstroPortal [bR⁺06] is an example grid application which enables the "stacking" analysis of multiple region images of the sky for predicting orbit changes and detecting faint objects. The basic paradigm of "stacking" is similar to a farm: the analyzed sky regions are subdivided by a master, the analysis is performed in parallel by the workers and the results are combined by the master into a solution. Different stacking functions are available to perform different kinds of analyses (e.g., statistical analysis or signal-to-noise improvement). In contrast to usual master/worker implementations (e.g., in the Farm-HOC), stacking allows each worker to apply one or multiple different analysis functions.

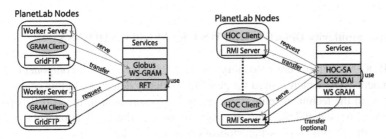

Figure 3.10 AstroPortal Setup for WS-GRAM (*left*) and HOC-SA (*right*)

AstroPortal originally uses WS-GRAM as its mechanism to transfer stacking functions specified by the user as worker code. The setup for AstroPortal is depicted in Fig. 3.10 (left). Each grid node in the PlanetLab environment [bP⁺02] can run both a GRAM client and a worker server. The requests issued by the clients carry the full application code (plus RSL) and are processed by one or multiple servers, depending on the amount of sky data. The transfer of input and output image files is handled using gridFTP and RFT [bM⁺02] of Globus, which is more efficient than SOAP [cW302] for this purpose.

Porting AstroPortal to the HOC-SA (as suggested in Fig. 3.10, right) would require a special "Stack-HOC" which takes into account the dynamic stacking aspect (individual functions per worker) of AstroPortal. Stacking is a coarse-grain pattern, i.e., the biggest part of work is performed within the multiple analysis functions, while the distribution of these functions among the workers is quite simple. Due to the fact that stacking is not a frequently used pattern, but rather specific for the AstroPortal application, the HOC-SA does not provide a "Stack-HOC."

Crash Test Simulations with HOC-SA

The Clayworks application is discussed in detail in the following chapter on large-scale applications of HOCs. Principally, Clayworks allows multiple users to construct objects (e.g., cars) collaboratively. Clayworks uses the Deformation-HOC from the HOC-SA component repository (Section 3.2.3) for simulating material deformations.

The Deformation-HOC is a multithreaded Java-based component that computes multiple simulation steps in parallel using multiple processors (typically, on one SMP machine. Results of experiments with different server types were presented at the eScience conference 2006 [bM+06]). If the processing of a single step were implemented as a self-contained program, WS-GRAM could be used to distribute the computations over multiple network nodes, as in AstroPortal. However, due to the fine grain of the computations in the Deformation-HOC, more parallelism will result in much more communication; therefore, the Deformation-HOC is implemented using the HOC-SA which is ideal for customizing only material-specific computation steps via user-defined code in the complex simulation.

Genome Similarity Detection with WS-GRAM and the HOC-SA

The HOC-SA component repository (Section 3.2.3) includes a component called Alignment-HOC which is used for genome similarity detection. In this chapter, the focus is on the technical setup, for running this component in the grid (Chapter 4 explains in more detail the *similarity matrices* used in this application and its differences to the genome processing application shown in Chapter 2).

Applications of the Alignment-HOC show that the HOC-SA and WS-GRAM are not always alternatives for handling the code transfer but both technologies are complementary and can be used in conjunction.

As any HOC, the Alignment-HOC is accessible via the Internet using a Web service. The parallel computations performed by the Alignment-HOC run on multiple networked servers which communicate via RMI. These RMI servers are neither dedicated grid nodes nor do they require any specific library or configuration, but any node can be used, i.e., the RMI server programs can also run on the same nodes as the HOC clients. The benefit of this setup is that input sequences can be processed locally on the client nodes reducing the data traffic. Each RMI server processes only a portion of data in this application but each RMI server is a self-contained program, i.e., it has its own main-method, and, thus, can be transferred via WS-GRAM to any node running Globus. The RMI server programs can be installed at runtime using WS-GRAM. Any grid node that has Globus installed can join a computation, upload its own input and automatically become a server that contributes to the distributed process, similarly to the well-known SETI@home project [bAn02].

Summary of the Case Studies

As demonstrated by the three case studies, which code transfer technology is the best – WS-GRAM or HOC-SA or both of them – depends on the application. The following three types of application design have been identified:

1. **Coarse-Grain Applications**: applications where all the data processing is centralised in large self-contained (*monolithic*) programs, rather than multiple components. Such applications are best transferred as a single block using WS-GRAM. The described AstroPortal fits into this category: the stacking functions executed by the users via this Portal are represented by executable programs which are simply transferred and run. (The described Stack-HOC would change the situation, but this hypothetical component was never implemented). Another example are legacy systems, which do not use components.

2. **Fine-Grain Applications**: applications where the data processing is distributed over multiple components separating the generic reusable parts from the problem-specific code. An example is Clayworks, based on the Deformation-HOC, which can compute deformation simulations for various materials. Users only specify how a single simulation step for a particular material type is computed while the largest part of the work is performed by the ready-built component. The steps which the users specify are not self-contained programs, but rather pieces of code uploaded (and exchanged) on demand using the Code Service. Obviously, the HOC-SA fairly reduces the data transfer costs in this design, since, when the application changes, only the affected operations must be transferred anew.

3. **Hybrid Applications**: Many large-scale applications (especially in distributed environments) require both components (with data and code parameters) and self-contained programs (e.g., server programs allowing to outsource parts of the computations) as well. An example are applications of the Alignment-HOC. These applications benefit from the HOC-SA allowing them to alter the computation of matrix elements via the code parameters for performing different kinds of genome analysis. At the same time, these applications can benefit from WS-GRAM, when it is used to install the server programs for processing parts of the data (submatrices) dynamically.

The above classification groups applications according to the relation between their generic parts (such as a farm implementation) and their problem-specific parts (single computational operations, e.g., a material deformation step). The size of the problem-specific code in a component-based application is also referred to as the computational grain of the application. From the above case studies, it can be concluded that HOC-SA better fits for handling the code transfer in fine-grain applications, as the largest part of these applications (the components) are installed, once and for all. WS-GRAM better serves coarse-grain applications, as all code transfer is handled in a single operation without the indirect way over the Code Service. However, in any coarse-grain application that has a generic part, a refactoring using components might be considered as it leads to better reusability. When only a part of the work in an application is performed using components, then WS-GRAM and the HOC-SA can both be used together in a single application.

3.3.3 Response Times: HOC-SA vs. WS-GRAM

In this section, the focus is on Web service availability and therefore, scalability is analyzed here in terms of the maximum number of possible concurrent clients. Performance experiments for different HOC applications with an increasing number of servers, as it is traditionally done in a scalability analysis, follow in Chapter 4.

The genome similarity detection application (Section 3.3.2) is used as a test application. Two implementations are examined: one on top of the HOC-SA and one on top of WS-GRAM (the previously mentioned version that uses both is not used in the comparison). The HOC-SA was deployed on grid nodes at the University of Chicago, with the following characteristics: dual-Intel Xeon^{TM}3.0 GHz with Hyper-Threading , 2 GB RAM and 100 MB/s network connectivity (for hosting the Code Service, the Web service for accessing the Alignment HOC and the Alignment HOC master). The (RMI-based) worker servers and the clients were distributed over the PlanetLab nodes [bP+02], which are connected to the PlanetLab overlay network with worldwide distribution. All nodes are Pentium III processors with processor speeds exceeding 1.0 GHz and most nodes are connected via 100 MB/s network links (although some still have 10 MB/s links). The operating system on all nodes is Linux (Kernel version > 2.6).

All the tests were conducted using the DiPerF [bD+04] tool in Globus, a tool for grid performance evaluation. For a precise analysis, the overall response time was split and the following metrics were studied in detail:

1. Queue Time (**QTime**): for HOC-SA, the time for dispatching a part of the genome data to an RMI server. For WS-GRAM, it is the time needed for submitting an RMI server program as a job;
2. Startup (**Startup**) represents the time to set up all elements of the application (i.e., initialize the input file transfer and launch the RMI servers);
3. Application Initialization (**AppInit**) represents the time required to have the application ready to start (all required files are in place and the RMI servers are ready to start the computation);
4. Time with Load (**xTWL**) represents the ratio between any of the time metrics above (**x**) and the number of concurrent clients (the **Load**); this metric is introduced for a response time comparison between services when loads differ;
5. Throughput (**Throughput**) quantifies the number of requests completed successfully by the service per time unit;
6. Throughput with Load (**TRWL**) is defined analogously to Time with Load.

Scalability Results for the HOC-SA

In the first set of experiments, 100 clients were mapped in a round-robin fashion on 100 PlanetLab nodes. A new client was created every 30 seconds and ran for 1800 seconds (lower part of the diagrams) or alternatively, every 180 seconds and it ran for 3600 seconds (upper part of the diagrams). During this time, each client repeatedly requested an alignment of random sequences. A total of 300 or 800 such alignment tasks were executed. This is the default testing setup for Web services in DiPerF [bD⁺04].

The HOC-SA performance with at most 20 clients is documented in the first series of measurements in Fig. 3.11 and with 40 clients running on 100 PlanetLab nodes in Fig. 3.12. Both cuves use the DiPerF-typical representation: the colored curves follow the values for the previously explained metrics: **QTime**, **Startup**, and **AppInit** on the left axis and **Load** and **Throughput** on the right axis. What metric is displayed in what color is documented in the legend and the colored arrows above the curves point at the vertical axis that belongs to the corresponding curve. The horizontal axis shows the time that has passed during the experiments. The curves show average values, while individual measurements are indicated by dots.

Comparison of the WS-GRAM and HOC-SA Results

We compared the response times of WS-GRAM with the response times of the HOC-SA: max. 20 concurrent clients (Fig. 3.13) and 40 clients (Fig. 3.14) running

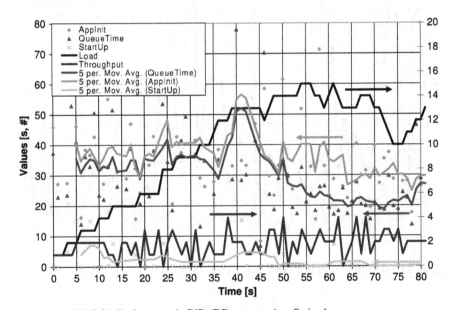

Figure 3.11 HOC-SA Performance in DiPerF Representation, Series 1

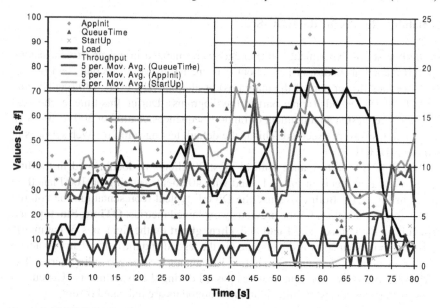

Figure 3.12 HOC-SA Performance in DiPerF Representation, Series 2

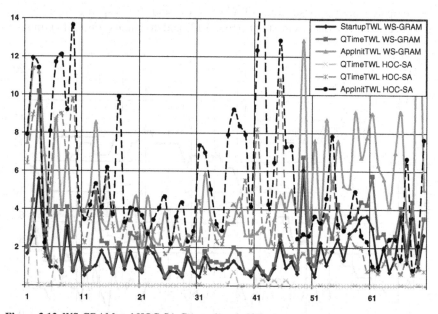

Figure 3.13 WS-GRAM and HOC-SA Comparison in DiPerF

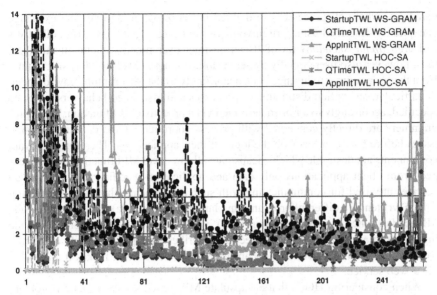

Figure 3.14 WS-GRAM and HOC-SA Comparison in DiPerF

on 100 nodes. In this diagram, only the **QTime**, **Startup**, and **AppInit** values are shown.

A throughput comparison has been conducted separately, leading to the result that the throughput mainly depends on the employed data transmission technique: when the large sequence files were transferred via gridFTP [bL⁺04] instead of using SOAP (incl. the time-consuming XML encoding), the throughput was up to three times higher, independently of running the WS-GRAM or the HOC-SA version.

The response time comparison shows that WS-GRAM delivers better startup times (up to two times higher performance, due to the caching of jobs). But WS-GRAM causes a CPU utilization that is up to 45% higher than in the HOC-SA experiments. Therefore, the performance of HOC-SA is more stable, when the number of concurrent clients rises. In the WS-GRAM experiments, also the memory usage was higher. Thus, the HOC-SA supports a larger number of concurrent clients. On average, the CPU utilization for the HOC-SA was by 20% smaller which can be crucial for the choice of the code transfer technology, especially if many applications run on the grid at the same time.

3.4 MPI, Skeletons and Web Services: Integrating Grid Technologies

The message passing interface (MPI [cAl08]) which was defined in 1994 is still arguably the most popular messaging library for HPC applications. MPI programs are written following the Single Program Multiple Data (SPMD [bFl72]) programming paradigm, where all processes are defined in one program and all data exchange is explicitly described using primitive send and recv functions. MPI collectives and *algorithmic skeletons* [aCo89] can abstract over process communication details, however also user-friendly skeleton libraries, e.g., *eSkel* [bCo04] still require from the programmer to arrange code accordingly to the MPI runtime environment, in particular, an explicit distinction of processes and associated behaviors has to be specified, accordingly to a *rank* determined via a primitive MPI function. Thus, programmers are directly concerned with process coordination and data transmission issues before they can focus on the logic of their applications. Therefore, software components are desirable which encapsulate all low-level MPI functionality and expect from client applications only data and application-specific code. When Web services are used for communicating with such components, the client application can be written in a different programming language (e.g., Java) than the server-side MPI code (which is usually written in C), while the client still benefits from the high-performance of C+MPI. This section describes an experiment with a skeleton from the *eSkel* library, which is encapsulated into a HOC that makes it accessible via a Web service.

When considering HOCs that encapsulate MPI-based skeletons, a new problem arises: How can the customization of skeletons (i.e., the description of application-specific code which is expressed using code parameters in the context of HOCs) be handled using Web services? In traditional MPI systems, client applications handle the customization of skeletons using the low-level mechanism of passing function pointers. These pointers refer to code units located in the same address space as the skeleton, which tightly couples the application to the library code. Moreover, the customizations specified this way affect only the sequential operations performing the application-specific work inside a skeleton, while the skeleton's parallel behavior remains fixed. The implementation of the skeleton's fixed parallel behavior is fine-tuned for the executing hardware (e.g., by running exactly as many concurrent processes as physical processors are actually provided). This design leads to good performance, but prevents programmers from applying component adaptations by spawning their own threads, as shown for the Farm-HOC in Section 2.4. If the most efficient solution of a certain problem requires changing the processing order belonging to a skeleton, the programmer can either engage himself in a library implementation as a less effective compromise or a new skeleton must be implemented for that particular purpose. Despite of the depicted difficulties, some time-critical applications are still inconceivable without C and MPI, since the latter offers the most direct way for coordinating processes and processors.

The central question studied in this section is: how can a framework of parallel software components plus a set of possible customizations provide a service-level abstraction over native implementation technologies?

3.4.1 A Gateway for Bridging between Web Services and MPI

In telecommunication terminology, a gateway is a piece of hardware or software that interconnects systems that use different protocols. This section presents a gateway bridging between SOAP and an MPI-environment for abstracting over inter-process communication. Technically, this gateway is composed of a specially configured Web service (called *gateway service*) which stores the MPI-related data (e.g., the number of processes) via the Globus-typical WS resource properties and a TCP server program which communicates with the MPI environment and the gateway service using a socket. This TCP server enables the gateway service to offer operations for launching and customizing MPI-based skeletons over the network. In the following, the gateway setup is explained in more detail. Although the example uses the *eSkel* library, the same setup can be reused in any grid application requiring MPI or another alternative efficient messaging implementation (for bypassing the heavy-weight SOAP encoding in processes communicating on the server side).

Contrary to a typical MPI-application, where the client itself is running on top of the MPI platform and coordinates the parallel computations as, e.g., process 0, a grid client accessing the gateway runs on a separate computer, as shown in Fig. 3.15. The Web service contacted by this client also does not run an MPI-process, but rather a thread on the Web service host (Globus) that forwards all input to the TCP gateway connecting it to the MPI processes. This gateway runs as one dedicated MPI process, which replaces the role of the MPI-client when the gateway is used, i.e., it coordinates other MPI processes running the skeleton code. From the viewpoint of the Web service-client, the parallel processes have no distinctive purposes, MPI communication is completely hidden, but the number of concurrently running processes can be adjusted via the gateway, thereby allowing the client to scale applications. As for all Web services, the communication of request and result data between the client and the Web service is handled by exchanging XML documents.

A particular feature, which distinguishes the presented gateway from other Web service systems is that it establishes a connection to an MPI-environment which is not running inside the Web service container and exhibits properties of its own. To maintain this connection, the gateway service must store some *state data*, i.e., data persisting the execution of single operations of the service: the number of the TCP-port used for transferring application data to the MPI-environment, the functions applied in a single stage of the MPI-based remote skeleton (a pipeline, in Fig. 3.15) and output variables for holding results. This state data is a part of the reusable configuration of any HOC connected to it (i.e., it is declared in its WSDL).

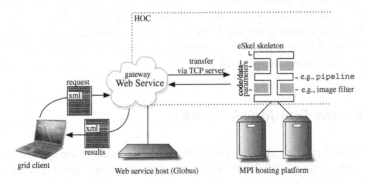

Figure 3.15 Connecting Different Hosting Environments via a Gateway

The separation of the MPI-environment from the Web service container and the use of an extra communication channel (the TCP server in the presented implementation) inside the gateway is a necessity for such a composition. Web service containers like Globus are parallel programs which can process multiple requests simultaneously via multithreading. Therefore, they must not be run within a multi-processing environment like MPI, which would lead to running one extra container per MPI-process, making resource sharing unfeasible and furthermore resulting in a process management overhead.

When a client invokes a HOC based on an MPI skeleton, the gateway service assembles a `command` string wherein a platform-dependent prefix holds the path to the MPI-installation directory, followed by `mpirun -np #` <*program name*> with the #-parameter reflecting the `numProcessors` resource property. Upon request of its `init`-operation, the gateway launches the MPI-program HOC by calling `system(command)`.

3.4.2 Example: Discrete Wavelet Transform (DWT)

DWT includes applications such as equalizing measurements, denoising graphics and data compression that must often be applied iteratively to large amounts of data. Therefore an efficient parallel implementation is desirable.

Despite the fact that some commonly used libraries for scientific computing, such as the GNU scientific library [cGS03], provide reusable DWT implementations, developing a special HOC for DWT makes sense for the following reason. While library procedures execute locally on the calling machine, a HOC that is remotely accessible via a Web Service allows clients to outsource the transform to a multiprocessor machine. In the grid context, where typically multiple Internet clients process data simultaneously, a standard desktop PC can quickly become overloaded with DWT computations, especially, when the transform is applied iteratively to large input data instances.

The result of DWT is reversible, i.e., the original input can be reconstructed using an inverse process, called wavelet synthesis. In an application, the transform is customized with parameter functions, such that the transformed data exhibits properties that can not so easily be detected in the source data, contours are accentuated or the transformed data can be represented using less memory. This customization is done by parameterizing a general schema with application-specific functions (implemented as HOC code parameters), making it an ideal candidate for using a HOC which runs the computations on the grid.

3.4.3 Wavelet Transform in General

Wavelet transforms are integral transforms, closely related to the (windowed) fast Fourier transform (*fft*). While *fft* decomposes a function into sines and cosines, the continuous wavelet transform is defined as

$$cwt(f;a,b) := \int_{-\infty}^{\infty} f(t)a^{-\frac{1}{2}}\overline{\psi}(a^{-1}(t-b))dt \tag{3.1}$$

Here, function $f(t)$ is projected onto a family of *wavelets* which are zero-mean functions (i.e., the domain average is zero), such as the second derivative function of the Gaussian probability density function. The graph of this wavelet is the commonly known "Mexican hat" shown in Fig. 3.16.

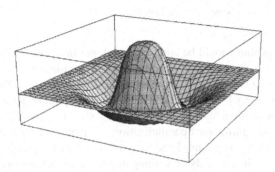

Figure 3.16 The Marr Wavelet, Sometimes Called the Mexican Hat

Instead of a continuous function, the discrete wavelet transform is a mapping of a set x of samples (such as a list or a matrix) that can be split into two equally sized subsets, u and v, each holding m elements. The "lifting technique" [bSw96] for computing DWT was discovered by W. Sweldens in 1994. Starting with all $u_{0,j} = x_j$, it transforms x by applying two functions called *predict* and *update* repeatedly, according to the following schema, called the *lifting scheme*, since each function application is said to lift the input toward a new step:

$$
\begin{aligned}
(u,v)_{i+1} &:= split(u_i) \\
u_{i+1,j} &:= u_{i,j} - predict(v_{i,j}) & \text{for } j < m \\
v_{i+1,j-m} &:= v_{i,j-m} + update(u_{i+1,j-m}) & \text{for } j \geq m
\end{aligned}
\tag{3.2}
$$

For each increment of i, j runs periodically from 0 to $2m$, thereby performing one *lifting step*, wherein the *predict* function is applied to the values in subset v ("predicting" subset u). The samples u_i are then replaced by the differences between their predicted values and their actual values. These differences are processed by the *update* function (as a "correction") and added to the samples in subset v. While the computation schema is fixed, the functions *split*, *predict* and *update* can be customized.

Implementation Schema of the Lifting-HOC

The wiring diagram of the two lifting steps in Fig. 3.17 graphically illustrates the structure of the lifting algorithm introduced formally in Section 3.4.3. The minus means that the input from the top is subtracted from the input from the left.

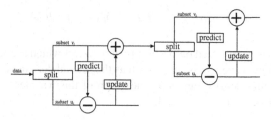

Figure 3.17 The Lifting Scheme, Forward Transform

When the algorithm should be applied to multiple independent data sets in parallel, the pipeline skeleton [aCo89] can be used for parallelization. The number of lifting steps that can be applied to an input set (called the *scale* of the transform in classical wavelet theory [aHu98]) is limited. If the first splitting produces subsets of m elements, the maximum number of steps is $log_2(m) + 1$, as the input is bisected in each step. For a straightforward parallelization, a pipeline wherein each stage corresponds to one lifting step is used and the total number of stages is determined by the largest input set. It can easily be verified that this basic setup works by reversing the schema: update and predict functions are swapped, such that updated values are subtracted and predicted ones are added as shown in Fig. 3.18.

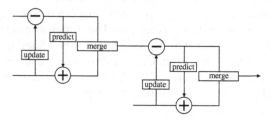

Figure 3.18 Data Syntheses Using the Backward Transform

A reverse pipeline with the same number of stages as the transform pipeline can be used for a synthesis to reconstruct the source data. In the following, an imaging application is presented which uses a component called *Lifting-HOC*: the Lifting-HOC combines the pipeline skeleton with a Web service using the gateway described in Section 3.4.1. The Lifting-HOC allows to pass a data array through the number of lifting stages corresponding to the array length, either in forward or reverse order, via distinct operations. The customizing *predict* and *update* functions can be passed to the Lifting-HOC as code parameters.

3.4.4 DWT for Image Processing

Distributed Web service applications often connect clients to image sources such as remote PCs equipped with sensors or cameras. In the following example, a demon-

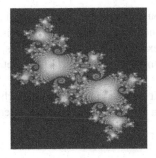

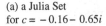

(a) a Julia Set
for $c = -0.16 - 0.65i$

(b) reconstruction
with threshold 2.5

Figure 3.19 Application of DWT on a Grayscale Fractal Image

strator setup is used, where one (or multiple) PC(s) generate(s) fractal images (such as the Julia set described in Chapter 2) and another PC (or set of PCs) filters them using the discrete wavelet transform.

Figure 3.19 shows the effect of the application of *dwt* on a fractal image. The input image is transformed up to the maximum scale (i.e., until no more splitting is possible, as explained in Section 3.4.3) and then reconstructed via the inverse transform. By setting all pixel values below a given threshold to zero in the transform, some details have been omitted in the reconstruction.

The fractal image used in this example features very fine contours that become bolder in the reconstruction. However, the original structure can still be recognized.

Contrary to number series that can be simply split into the even and odd entries, images require a customization of the split function that specifies an adapted 2-dimensional partitioning. If a program simply concatenated all rows or all columns of the image matrix into an array, the image structure would get lost during the transform, as most neighboring entries in the matrix are disjoined in such an array. Instead, the presented implementation overlays the image with a lattice that classifies the pixels into two complementary partition matrices which preserve the data correlations.

3.4.5 DWT on the Grid Using the Lifting-HOC

This section shows how to use the Lifting-HOC for computing DWT on the grid, thus allowing the client to customize the transform remotely via Web service parameters. The behavior of the DWT implementation in the Lifting-HOC is based on the lifting scheme (introduced in Section 3.4.3) and essentially depends on its predict and update functions.

The split function that was employed in the image processing example in Section 3.4.4 creates a so-called *quincunx* lattice: All pixels are alternately assigned to a subgroup of black pixels or to another subgroup of white pixels arranged like on a chessboard, i.e., the color pattern is shifted by one pixel in each row. The quincunx is just one possible lattice among others, e.g., hexagonal or octagonal lattices. The implementation of the split function is not shown here (see [aJH01] for a detailed description).

The Lifting-HOC allows to parallelize wavelet transform applications, such that the stages run simultaneously when multiple data are processed. The Java interface of the Lifting-HOC reads as follows:

```
1: public interface LiftingHOC<E>  {
2:   public void configure(String [] settings);
3:   public E[] process(E[] data);
4:   public void  predictDefinition(Function<E> f);
5:   public void  updateDefinition(Function<E> f); }
```

Figure 3.20 Java Version of the Lifting-HOC Interface

Line 2 defines the `configure` method which takes the `settings` parameter, encoded as a `String` array. As usual for HOCs (see Chapter 2), this parameter is used to specify the names of grid servers which will be assigned to perform a portion of the computations. A possible alternative is to use the `settings` parameter for specifying a file wherein a mapping between servers and lifting steps is specified, such that dedicated servers are used for one or multiple steps.

Line 3 is used to trigger the processing of `data` which has a variable type (therefore the type parameter E). To ensure that all data is exchangeable among all hosts, the type range for E is limited to primitives and collections of primitives (a differently user-defined `class` type will cause a serialization error).

Lines 4–5 define the methods for specifying the stage functions `predict` and `update` which are non-primitive, but must be instantiated with a user-defined class holding executable code. The contents expected in the definition of f are defined in the `Function`-interface forming a contract between the HOC implementer and the user of the HOC. Thus, the parameter f of the Lifting-HOC is defined by another interface, named `Function`, which must be implemented by the stage functions.

In the example application from Section 3.4.4, an instance `lh` of the Lifting-HOC is used for applying a filter to images. The `Function` parameter f in such an application of DWT can be defined in Java as follows:

```
1: lh.configure(new String[] {"P0", "P1"})
2: lh.predictDefinition( new Function<int[][]>( )  {
3:    public int[][] f(int[][] m) {
4:       return imageFilter(m, threshold); } } );
```

Figure 3.21 Using an Instance of the Lifting-HOC in a Java Program

Line 1 passes the processor names P_0 and P_1 in an array and thereby defines a mapping between the processors and the lifting definitions starting in line 2. The `imageFilter` method, used in line 4 for the predict definition, takes two arguments. The first argument m is the image matrix and the second is a value defining what data should be removed from the matrix m after computing a prediction. Figure 3.22 shows the effect of calling `lh.process( )`.

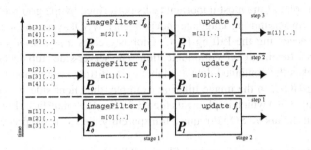

Figure 3.22 3 Lifting Steps of a 2-Stage DWT Application on 2 Processors

It can be seen that the DWT application is executed by the Lifting-HOC via pipelining: in step 1, processor P_0 filters the row `m[0][..]` while processor P_1 stays idle. In step 2, P_1 updates row `m[0][..]`, while P_0 filters row `m[1][..]`. In step 3, the result of filtering row `m[0][..]` is available, such that P_0 and P_1 can start processing rows `m[1][..]` and `m[2][..]` in parallel.

Accessing the Lifting-HOC Remotely

SOAP is used for exchanging code and data between the Java client and the server-sided C implementation of the Lifting-HOC, i.e., program entities are encoded into XML-elements which are placed in XML-documents communicated over the network.

```
1: <soap:envelope><soap:body>
2:   <liftingHOC:process>
3:     <liftingHOC:item>2</liftingHOC:item>
4:     <liftingHOC:item>3</liftingHOC:item>
5:     ... <!-- more items -->
6:   </liftingHOC:process>
7: </soap:body></soap:envelope>
```

Figure 3.23 Transfer of the Data Parameters in the SOAP Format

Figure 3.23 shows, e.g., the (shortened) SOAP request that must be transferred, when the user wants to process the number series { 2, 3, ... }, and, for this purpose, calls `lh.process(new double[] { 2, 3, ... })`. The data being passed is placed in the lines 3–5 of the above code (12 lines have been omitted;

they contain additional type information for binding the soap and liftingHOC-prefices – both identifiers are arbitrary – to XML-namespaces plus the definition of the array holding the parameters and some declarations regarding the HTTP-encoding of the request). Such a SOAP request is typically not written manually. Automating the decoding and encoding of such documents and transmitting them over the network is one of the most important tasks performed by the grid middleware.

From the (programming-language-independent) WSDL-version of the remote interface for accessing the Lifting-HOC, Globus creates a server-side stub and a client-side stub for the SOAP communication. The server-side stub decodes the request and makes the contained data available to the lifting scheme implementation in the Lifting-HOC. In the image filtering example, the client-side stub is the Java object lh (Fig. 3.21) which implements the LiftingHOC interface (Fig. 3.20) and allows to emit the above SOAP-request transparently, i.e., as if the processing took place locally.

A HOC must contain configuration files in addition to code, due to two distinctive features of its remote interface:

(1) the methods in the component become operations of a Web service which is not object-oriented, i.e., there is no implicit connection between the operation code and data objects or other program entities. Therefore, the HOC developer must explicitly configure WS resources for representing state data, i.e., data records that are connected to the operations of a Web service [cOA04];

(2) code-carrying parameters must be mapped onto primitives by the middleware, as there is no corresponding type for a parameter requiring its own interface in WSDL (e.g., parameter f of the Lifting-HOC, which requires the Function-interface and is declared via a primitive identifier in the configuration, as shown in Section 3.4.5).

To exchange code in the HOC-SA, the Code Service was developed. The idea behind this service is to make the most common interfaces (e.g., Function) for code parameters publicly available in a table which connects each interface to a primitive identifier. Thus, a remote user can refer to an interface by passing its identifier to a HOC, which can then look it up using the Code Service and use it for accessing a code parameter, which is always transferred in a raw binary format itself.

In C+MPI-based programs, code is referred to by passing function pointers and called by jumping directly to its first instruction, instead of following an interface definition. When a HOC makes use of a C code, as the Lifting-HOC does, there is no need to define how these parameters can be accessed or executed. For this reason, only one specific identifier is used for marking any C code in the Code Service indicating that there is no need for an explicit interface definition.

3.4.6 Portable Parameters for the Lifting-HOC

The Lifting-HOC has a public interface by means of which the application-specific *predict* and *update* functions can be uploaded as strings in order to customize the Lifting-HOC for a particular wavelet application. The pipeline stages are predefined,

such that the flow of operations within one stage exactly adheres to the wiring plan
of a lifting step shown in Section 3.4.2 and the operations can be customized.

In an implementation of *dwt* via lifting, not only *predict* and *update* but also
the *split* function is usually an application-specific parameter. Due to the fact that a
String representation of arbitrary *split* functions is difficult to define (one possibility would be a predicate term checking whether some input belongs to the first
or second set), the splitting of data is a fixed part in the Lifting-HOC. The HOC
provides the most common splitting procedures: In the case of one-dimensional input, the Lifting-HOC splits the data into the elements with even and the elements
with odd indices. In the case of an input matrix, like in the imaging application from
Section 3.4.4, the *split* function computes the so-called *quincunx lattice*, i.e., all the
pixels of the processed image are alternately assigned to a subgroup of black pixels
or white pixels. To compute the image transform shown in Section 3.4.4, the custom
predict function rates the grayscale value of a pixel by computing the average of its
nearest neighbors:

$$predict(x_{i,j}) = \frac{1}{4}(x_{i-1,j} + x_{i,j-1} + x_{i+1,j} + x_{i,j+1})$$

The corresponding *update* function returns half of the average computed by the
predict function:

$$update(x_{i,j}) = \frac{1}{8}(x_{i-1,j} + x_{i,j-1} + x_{i+1,j} + x_{i,j+1})$$

This updating procedure reflects the bisection performed by the *split* function in
each lifting step: it preserves the average grayscale value of the input during lifting,
i.e., the value average over all pixels in both partitions equals half the average over
all initial values. Some more sophisticated methods also use neighboring values, but
with a different calculation rule. In most cases, both *predict* and *update* are one-step
procedures. Therefore, most *predict* and *update* functions used in wavelet lifting can
be defined using simple assignment terms (without loops, conditional statements or
other complex control structures). The Lifting HOC accepts these parameters in an
alternative format: instead of implementing a Java interface (such as Function,
shown in Section 3.4.5) or sending these procedures as C programs, the terms can
be directly transferred to the server side and interpreted there. Implications of using
either Java, C or interpreted string terms for encoding code parameters are discussed
in Section 3.4.9.

3.4.7 An Adaptation of the Lifting-HOC

In Section 2.4, component adaptations were introduced, to adapt the Java-based
Farm-HOC to the wavefront pattern for increasing the level of parallelism in the
sequence alignment application. This section presents a possibility for adapting the
Lifting-HOC (based on the *eSkel* pipeline) for more efficient parallel processing.

The difference between the data being processed by the lifting algorithm as compared to many other pipeline applications, is that the lifting algorithm works with data sets of varying sizes and that the number of data elements that are affected by a stage is reduced in each stage. The *eSkel* pipeline can be made especially suitable for this type of application. The presented adaptation of the Lifting-HOC is static, i.e., the user is not required to run any specific adaptation thread (as in Section 2.4) for using the adapted component.

The Stage-Skipping Optimization

The following optimization can be applied to any pipeline-based implementation of DWT, but this section shows how it is implemented in *eSkel* where different *interaction modes* help the user in implementing optimizations. A special feature of the lifting algorithm is that the data being processed is reduced during each step. The stage-skipping optimization makes use of this fact to shortcut several steps, when running the lifting algorithm in parallel. Indeed, the several inputs of the pipeline may be of varying sizes, and the number of lifting steps is directly related to their size. Hence, an input of short size does not need to go through all the stages of the pipeline, as it happens in the parallelization described in Section 3.4.3.

In *eSkel*, variations of the standard behavior of library functions are defined using so-called *explicit interactions* [bG⁺06] between activities (i.e., the communication of data which is used as input or output of a code parameter). These interactions express non-standard behaviors and release temporal constraints.

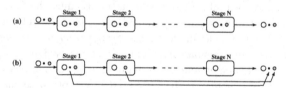

Figure 3.24 (a) Standard *eSkel* Pipeline Skeleton; (b) Stage Skipping Optimization

Figure 3.24(a) displays the standard pipeline behavior with implicit interactions: each data that is produced by a stage function (e.g., by a user-defined *predict* or *update* function) is passed to the next stage function by the skeleton (i.e., implicitly). The circles represent three examples of input items, the size of the circles is proportional to the size of the input.

Figure 3.24(b) illustrates the stage skipping optimization: the smallest input is finished in stage 1 and no more present in stage 2 (while it had to be towed through all stages in Fig. 3.24(a)).

In a HOC code parameter, the user can explicitly write the code for communicating the input and output data of a stage function, such that, e.g., stages can be skipped. In the example of the *eSkel*-based Lifting-HOC, the stage skipping is implemented using the explicit interaction mode: instead of retrieving all its input for a stage function through its parameters and passing all output as a return values, the user directly calls the *eSkel* functions `Give` and `Take`. Using these functions, any stage can send output not only to the next stage but also to the last.

3.4.8 Experimental Performance Evaluation

The focus of the following brief performance analysis is on the implications of using the gateway from Section 3.4.1 in the Lifting HOC, or, more general, on the implications on the performance of applications, when Web services are connected to any C+MPI-based software. Primarily, the following two questions are studied:

(1) How much time overhead is introduced by the use of Web services, i.e., SOAP communication as compared to running all computations locally?
(2) Does the introduction of the additional indirection ((TCP), required for the MPI gateway from Section 3.4.1) seriously impact data transmission times?

In the experiments, two one-dimensional input sets were used as input for the DWT example: a *small series* that contained 64 independent lists of 4096 `double`-precision numbers each, and a *large series* that contained 64 lists of 16384 `double`-precision numbers each.

	SOAP Time	MPI Gateway Time (TCP)
small series	3370ms	3729ms
large series	12165ms	5782ms

Figure 3.25 SOAP & MPI Gateway Times of the DWT Example in a LAN

During the transmission of the number series, the portion of time spent for the SOAP encoding was measured. The time spent for the TCP transmission inside the gateway was measured separately. The table in Fig. 3.25 shows the average values for 3 repeated measurements using only machines in a local fast-Ethernet network with a latency below 1 ms.

All measurements were performed also in a wide-area scenario: here, the distance between the client and the Web service was approximately 500 km resulting in a network latency above 20 ms. The gateway and the MPI environment still resided inside the same LAN, thus, the TCP transmission times are almost the same (for the small series, the average TCP transmission time was even lower than in the LAN, but only by 20 ms, which documents that there were low deviations among the repeated measurements). Table in Fig. 3.26 shows the results for these tests.

	SOAP Time	MPI Gateway Time (TCP)
small series	5844ms	3702ms
large series	16849ms	5843ms

Figure 3.26 SOAP & MPI Gateway Times of the DWT Example in a WAN

The experimental results help answering the first of the above questions: Web services introduce a considerable amount of time overhead, especially in data-intensive applications (also in a LAN). The characteristics of the network (latency) do not heavily impact the data transfer costs, which can be explained by the high costs of the SOAP encoding process. Therefore, connecting, e.g., a Java program to an MPI-based system using Web services and the described gateway is only reasonable if the computation time spent in the remote MPI-system is much higher than the required transmission time (ranging from some seconds up to several minutes). In this case, it is possible to speed up the overall processing time, when the employed MPI-based system (e.g., a cluster) can run the computations much faster than it is possible locally. In the DWT application, this decision requires (besides the presented transmission time measurements) a scalability analysis of the parallel wavelet transform computation itself. For the Lifting-HOC, as presented in this book, such an analysis has not been conducted (a more extensive performance analysis of the lifting algorithm on top of MPI can be found, e.g., in [bG$^+$01]). However, when an efficient C+MPI-based implementation of the wavelet transform is provided, a Java programs that connects to it (instead of running all computations itself) can gain better performance, even in a LAN.

Concerning the second question from above, the measurements prove that the additional TCP times (introduced by the MPI gateway) have a justifiable impact on the overall transmission times only if an application is data-intensive (e.g., the large series test). For the small series, in contrast, the gateway almost doubled the transfer costs and even the pure SOAP costs of multiple seconds are out of scale for an application which only filters a few thousand numbers. Due to the high data transmission costs, applications which only quickly process small data sets deliver better performance locally than on a grid platform with Web service-based middleware, regardless of the quantity and power of the grid's processing nodes. Applications with a high computation costs per data element are much better suited for this platform type.

As an experiment with an application that has higher computation costs (as compared to transforming number series), a 1024x1024 pixel true-color image was filtered multiple times, using DWT as described in Section 3.4.4. In this application, the overall transfer time was 12150 ms in the LAN and 19849 ms in the WAN scenario. Thus, transferring the image over the WAN takes only a little more time than in the LAN, while (depending on the remote servers) a lot of time can be saved when the processing is performed remotely.

3.4.9 Discussion: Interoperability and Portable Code

Although the Lifting-HOC, introduced in this chapter, was implemented using MPI, its code parameters, used for the application-specific customizations in Section 3.4.5, were all specified without calls to functions from a particular MPI library. However, if the customizations are represented using C-Code, there is still the problem that pointers to local code (representing code parameters in a C program) cannot be passed across network boundaries. In a client program that is implemented in another language than C, moreover, the question arises, what format should be used for encoding the code parameters. In this section, possibilities for exchanging code over the network in a distributed system which is (partially) implemented in C are discussed in more detail.

A Machine-Oriented Implementation of Code Mobility

The problem of transferring a single C function can be addressed by a *code cut-out-procedure* that copies the function body into an array which can be posted across the network (using SOAP or other protocols). The reverse conversion (from an array to a function) is possible via a simple typecast. For this book, a code cut-out-procedure using such conversions was implemented and experimentally integrated into the Lifting-HOC.

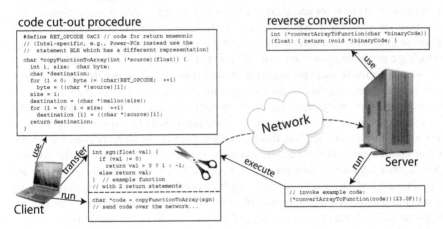

Figure 3.27 Example C Program that Cuts Out a Piece of Its Own Code and Transfers It Over the Network for Execution on a Remote Server

As schematically shown in Fig. 3.27, the cut-out procedure finds the correct part to be cut out of a local program by determining the boundaries of the C functions to be sent over the network. The beginning is given by the address of the function that is cut out (sgn, shown in the box with the scissors symbol in the depicted example). This address is specified by the client program via a pointer, while the end of the function is found by copyFunctionToArray which scans the binary code starting at this address for the assembly language representation of the return statement (the RET_OPCODE, assumed that this is the last statement in the binary code of the function).

In experiments with the GNU C compiler [cGC05], the described code cut-out procedure worked on Intel and Power PC processors, since with this compiler, the above assumption holds: it produces exactly one return statement for any subroutine (functions that contain more than one return statement in the C source, as sgn in Fig. 3.27, are translated into Assembler using conditional jump instructions). Unfortunately, features like this can only be observed via "hacking" and such observations cannot be generalized for any C compiler. Moreover, cutting out pieces of C programs and exchanging them across the network is non-portable across CPUs with different instruction sets. However, the exchange of C functions as code parameters using the presented cut-out procedure facilitates the adaptation of C-based HOC (e.g., for the stage-skipping optimization shown in Section 3.4.7), since it allows the user to directly interact with the component code.

A High-Level Implementation of Code Mobility

By reviewing the customizations from Section 3.4.6 carefully, it can be seen that the *predict* and *update* functions, as well as the stage skipping criterion in the optimization from Section 3.4.7, can be represented using any programming language that supports arithmetics and array access. Implementing a parser and a compiler for a simple term language, which supports only these constructs is not more difficult than the most basic examples for the popular compiler construction tools lex & yacc which are available open-source in different C/C++ and Java versions [aMB90].

We conducted some experiments with the Lifting-HOC, where the code parameters were exchanged using an interpreted term language, as a more high-level alternative to the previously presented code cut-out procedure. In these experiments, the term language used for representing the code parameters was specified by the user via a grammar parameter. Via this parameter, the Lifting-HOC becomes a "third-order component" (TOC), i.e., the grammar specifies the language of the component's other parameters (the *predict* and *update* functions). Interpreted strings are the simplest format for representing portable code for any component that supports a specific parameter language, regardless of the employed interpreter. The JEP interpreter [cSi06], e.g., can be employed by a component which supports parameters written in a fixed term language [cSi06]). A TOC which allows clients to upload code parameters in a user-defined language will rather employ an expression parser based on lex & yacc (which also handle the required type checking) and expect a BNF grammar as an additional parameter.

String interpretation is the preferable technique for exchanging code in any application like DWT, where all code parameters can be expressed using terms (i.e., without flow control statements like, e.g., loops). Especially when the client is Web-based and allows the user to directly enter the code parameters (instead of only selecting them from a pre-built remote repository). When the Lifting-HOC from this section is used to filter images remotely, a portable format for defining the splitting lattice (e.g., the quincunx, explained in Section 3.4.5) might be a semi-structured description of graphs (e.g., adjacency matrices), allowing users to modify the *split*-function, required for the 2-dimensional transform, in a similar manner as *predict* and *update*. The option of defining specific grammars for interpreting code parameters opens perspectives of designing TOCs for new classes of applications, e.g., a parallel string processing TOC that evaluates text replacement rules specified via a grammar, as they are used in the interpretation of L-Systems in algorithmic botany [aPL91].

From the experiments with the Lifting-HOC, it can be concluded that the concept of HOCs which can be parameterized by data and code as well is a promising approach toward abstracting over MPI-like primitives in higher-level components. The DWT example has also shown that users can easily modify the way both types of parameters are processed by the HOCs and, thereby, make the provided HOCs more suitable for their applications: (1) the programming language and format used for representing HOC code parameters can be defined by the users; (2) the control flow for processing the data parameters of a HOC can be modified via adaptations.

3.5 A HOC-SA Based Map/Reduce Implementation

Map/Reduce is a programming construct that is often mentioned in the context of future-trend distributed architectures, in particular *Cloud Computing*. Cloud computing is the most recent grid-related concept that made its way into the mass media [bLo07] and triggered considerable investments of major computer companies [cFe08]. Since more and more important technology providers, including *Amazon, Oracle* and *Yahoo!* agree on the *Cloud* as a new platform that brings all recent grid computing trends together, we consider combining the HOC-SA with the Cloud's currently most popular programming paradigm: Map/Reduce.

This section briefly sums up the characteristics of recent Cloud technologies, such as *virtualization* and *Utility Computing* [cWi08]. Then we discuss their relevance for HOC-SA based Cloud Computing systems. Finally, we outline a current Map/Reduce implementation for Cloud Computing and show how it can benefit from the HOC-SA.

3.5.1 Cloud Computing Technologies for the HOC-SA

Cloud computing [cGr08] is proposed to unify various grid computing approaches, in the area of both virtualization and Utility-Computing as well as other recent Internet-related topics, such as the Web 2.0 [cOR05].

Virtualization of Resources

Modern grid middleware is providing user-transparency to the extent that applications are built in terms of logical services, rather than by binding them to physical resources. This approach is sometimes called *virtualization* of resources. The mapping between services and resources is specified in the middleware configuration. Typical applications are scientific simulations, e.g., in astronomy, genetics or engineering, with very high demands concerning memory, computing power etc.

Utility Computing

Utility Computing is rather targeted towards the average Internet user or enterprise, than towards experts and scientists. It includes applications such as Google Apps (google.com/a) and Salesforce (salesforce.com).

The idea is that computational resources (e.g., storage) are rented instead of bought, and office or business applications (e.g., for processing mail, documents or customer requests) are installed on Internet-hosts providing such *Software as a Service* (SaaS). One obvious advantage over traditional software is that users only pay for as much of the remote resources as they actually consume. Moreover, users are freed from software maintenance (updates and backups) and a broad range of possible client platforms can be used, including laptops and embedded or mobile devices like smartphones, etc. Utility Computing is closely related to grid computing with respect to the resource virtualization aspect explained above; it often uses the WSDM framework which is quite similar to WSRF in the Globus middleware.

Cloud Computing

Cloud Computing brings together recent Internet technologies in one "Cloud," such that any *outsourcing* of program tasks over an Internet connection (including standard Web service invocations) can be viewed as an example application [cGr08].

In Cloud Computing, mission-critical jobs (incl. real-time computations) and standard application activities (e.g., rendering a graphical representation of some data) become tasks for the same type of platform. Interestingly, the Cloud Computing implementations from major vendors, such as IBM's *blue cloud*, rely on large compositions of equally equipped computers running the same software, e.g., *Hadoop*, see below. Thus, Cloud Computing still hides from the user the localities of the compute nodes (typically, behind a Web service), but heterogeneity among the remote compute nodes is not as explicit as in grid computing systems. This setup allows programmers to run a legacy system "behind the cloud" instead of porting everything to a new, special Cloud Computing middleware.

If one imagines the future of computing as a world where today's laptops, desktop PCs and supercomputers are all replaced by thin terminals inside TVs or phones which connect to some Internet-based "cloud" for running any kind of task, whether it is reading an email or conducting a complex physical experiment, then a uniform programming model for *cloud tasks* becomes desirable.

3.5.2 MapReduce and Hadoop

Many applications share a Map/Reduce structure [bDS04]: input key/value pairs are *mapped* to intermediate key/values via a user-defined *map* function, specifying *map tasks*. All intermediate values with the same key are reduced via a user-defined *reduce* function, specifying *reduce tasks*. With these two types of tasks, a broad variety of applications can be programmed.

An example application for Map/Reduce is a text analysis task like this: determine how many times a certain word or phrase is contained in a large composition of texts. For this purpose, the *map tasks* run a *map* function that leaves any input element unchanged, while 1 is returned as the corresponding intermediate value. After the *reduce tasks* for all data are executed, such that a *reduce* function sums up all intermediate values, the result is a list of all unique input elements plus their number of occurrences.

A similar algorithm is used for rating Web search results inside the popular search engines Google and Yahoo. If such a text analysis should be implemented for the grid using HOCs, we would need a Map/Reduce-HOC which runs the *map* and *reduce tasks* in parallel using code parameters which define the required *map* and *reduce* functions. However, modern cloud computing frameworks like Hadoop already provide such a component and no new HOC has to be developed for this purpose. Let us use Hadoop as an example for discussing how Map/Reduce can benefit from the HOC-SA.

Hadoop is the Java-based open-source implementation of Map/Reduce available from `http://hadoop.apache.org`. Many different companies including, e.g., *Amazon*, run Hadoop. *Yahoo!* claims their installation on over 10,000 Linux machines to be the world's largest Hadoop production application [cYD08]. The core of Hadoop is a Web service-based submission engine for *map/reduce tasks*, i.e., user-defined *map* and *reduce* functions in Java notation. This engine includes support for efficient scheduling, such that machines storing most of the data also handle most of the jobs, which keeps the network traffic low. Besides its own *Hadoop File System* (HDFS), Hadoop supports processing of distributed data which is stored using Amazon's *Elastic Cloud*, one of the first commercial implementations of hirable computing power for cloud computing [aMu08].

3.5.3 HOC-SA Features for Map/Reduce on the Grid

In the Map/Reduce model explained above, the master-worker communication is a part of a common framework implementation, e.g., Hadoop. In Cloud Computing, clients which are external to the cloud must be enabled to specify a part of the application code (the *map* and *reduce* functions).

When a Web service is used for this purpose, HOC-SA Code Service can be used for enabling the code transfer. Especially sharing of code among multiple workers in a Hadoop system can be implemented quite efficiently using the HOC-SA Code

Service: The Hadoop engine (see above) takes as its parameters Java implementations of *map* and *reduce* functions, which are submitted to so-called *Job Trackers*. Therefore, in HOC-SA terminology, Hadoop *Job Trackers* are a kind of HOC, i.e., a software component which allows clients to supply it with some application-specific code that is executed in a broader context (a parallel implementation of Map/Reduce in the case of Hadoop). Hadoop *Job Trackers* are known as critical points of failure, i.e., failed nodes can cause the loss of jobs and require new submissions of code. When the HOC-SA Code Service is used for storing the code on the cloud-side of the network, it can be shared by all workers, leading to a loose coupling between all cloud workers.

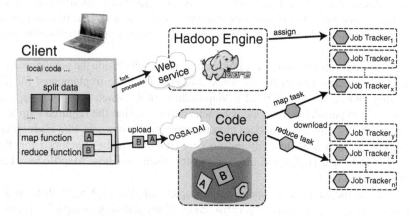

Figure 3.28 Possible Combination of the HOC-SA Code Service with Apache Hadoop

This idea is sketched in Fig. 3.28: the client uploads the *map* and *reduce* functions to the code service using OGSA-DAI, as for any other code parameter. Then, after executing the code which initializes the data locally, it splits the data and requests from the Hadoop Engine to fork a new process for every data item. These processes assign the data to the available Job Trackers which independently download the corresponding map or reduce tasks (depending on the application phase) from the code service and run them. This means that code never needs to be resubmitted, as it is the case in the current Hadoop implementation from Apache.

Once an application gets more complex than it can be expressed via a single application of the Map/Reduce pattern, other software components than Hadoop become useful. The HOC-SA includes many different kinds of HOCs, which all take code and data as their input that is processed in parallel using networked compute nodes (e.g., in the "cloud"). These HOCs are similar to Hadoop's Job Trackers, however they all offer different processing schemata such as pipeline, or farm processing, rather than only the Map/Reduce schema. Some HOCs even offer specialized processing schemata for specific types of data- and compute-intensive applications, e.g., genome string analysis or image processing. If an application uses multiple HOCs, or HOCs plus Hadoop, workflow descriptions become useful [bA+06], as they allow programmers to describe repeatedly used compositions of distributed computation patterns.

3.6 Summary of HOC-SA Features

Probably, the most important feature of the HOC-SA is its novel code transfer technique for handling mobile code in systems that use Web services for communicating among clients and components (HOCs).

Component software, as presented in Szypersky's book [aSz98], focuses on compositionality, which is also the main feature of the CoreGRID Grid Component Model (GCM [cCN05], see Chapter 3). The GCM and the SOA design of the HOC-SA are complementary. HOC-SA neither implies its own concept for object composition nor defines a linkage to a proprietary component implementation technology; SOA rather focuses on communication issues. Thus, the HOC-SA Code Service and the Remote Code Loader are valuable add-ons to the CoreGRID GCM.

The novelty of HOC-SA is that it merges service-orientation with the traditional component idea. The need for this new methodology is driven by the fact that components for the grid must be both, composable at the high level of application logics and interoperable via message exchange: HOCs can be assembled from services and other software, e.g., RMI/CORBA systems (Chapter 1), multiple different Web services (Chapter 2), or MPI programs (Chapter 3).

The support of script languages for parameterizing HOCs corresponds to the trend of Web application development, where, e.g., the Ruby On Rails (RoR) [cRR07] framework has gained a lot of popularity recently.

This chapter illustrates that many applications can achieve flexibility from the HOC-SA, allowing them to exchange the single operations performed by remote components using the Code Service and the Remote Code Loader. The management of the code units implementing the exchangeable operations is handled via OGSA-DAI which fosters code reuse, i.e., the same code units do not need to be transferred again and again. Besides code reuse, OGSA-DAI has the following advantages for the management of mobile code: transparent grid-wide data access, use of different data discovery strategies, and enhanced security. For executing native code in a HOC (as in the DWT application in Section 3.4.2), the code table in the Code Service has an extra column that is used to describe the code type. This type can determine, e.g., the required target platform for running a C program or a BNF for a user-defined language, leading to a high degree of flexibility.

Code mobility is also exploited by mobile agent systems, where units of code "roam the network," i.e., distribute and redeploy themselves onto remote computers. In contrast to code parameters in the HOC-SA, the behavioral patterns of mobile agents are inherent features of the agents, rather than implied by a target component. Typically, mobile agents are built using technologies that include an off-the-shelf code transfer mechanism (e.g., the *valuetype* construct in CORBA [cCO04]). The use of mobile agents in the grid has been studied, e.g., in [bS+02].

To cope with various different database types, OGSA-DAI is more appropriate than, e.g., an O/R-mapper like *hibernate* [cJB07] for grid applications, since O/R-mappers can only map all data to the tables of relational databases. The computations performed in grid applications are far more complicated than that of a typical business application. One difficulty is, e.g., the load balancing: vast amounts of concurrently used shopping carts can easily be handled, as these components have no dependences between each other. The container can automatically outsource instances of such cart components, as soon as the customer quantity exceeds a certain limit (defined in the container configuration). Contrary, computational work units that can emerge dynamically, e.g., during the evaluation of a divide-and-conquer algorithm, have causal and temporal dependences and can therefore not be handled as easily. HOCs have been developed to cope with such problems, providing a component architecture appropriate to the demands of distributed computational applications.

Finally, the relation between HOC-SA and WS-GRAM was clarified in this chapter. Different applications have different requirements, and coarse-grain applications are still better handled using WS-GRAM than using the HOC-SA. However, the time costs of most grid applications are not caused by the complexity of the single computations but rather by pure masses of computations and the data being processed. A HOC that provides a distributed processing schema, can speed up such applications significantly by increasing the number of execution nodes. The following chapter presents some real-world application examples using HOCs.

The full HOC-SA project (including source code, documentation, example HOC implementations and applications) is freely available as an optional extension to the core Globus toolkit (a so-called Globus incubator project [cD⁺06]) for downloading from the Globus Web site.

Chapter 4
Applications of Higher-Order Components

This chapter shows how HOCs are used in large-scale applications that require multiple computational resources. The introductory examples on HOCs in Chapter 2 and the HOC-Service Architecture (HOC-SA) in Chapter 3 were concerned with processing number series, string data and images or matrices, following quite simple rules, and the computational load was only created synthetically by copying input data for increasing its size and repeatedly running the same processes.

Contrary, this chapter shows two applications that tackle real-world problems and follow much more complex computational patterns:

(1) The first application, shown in Section 4.1, called *Clayworks*, deals with the simulation of the deformations resulting from the crash of clay objects moving in a virtual space [bG+09]. Clayworks is a software system which integrates collaborative real-time modeling and distributed computing. It addresses the challenge of developing a collaborative workspace with a seamless access to high-performance servers. Clayworks allows users to model virtual clay objects and to run compute-intensive deformation simulations for such objects crashing into each other. To integrate heterogeneous computational resources, modern grid middleware is adopted and users are provided with an intuitive graphical interface. The parallel computation of simulations is done using a specific, new HOC, called *Deformation-HOC*, as it can compute deformations of different kinds of material, expecting as its input objects, movement vectors, velocities and material characteristics. Clayworks is a representative of a large class of demanding systems which combine collaborative modeling with performance-critical computations, e.g., in engineering applications (car crash tests), biological evolution, or geophysical simulations.

(2) The second application, described in Section 4.2, deals with genome string analysis. Section 2.4 already introduced the basic string alignment computation as an example for adapting HOCs via code parameters. The application shown in this chapter makes use of the alignment computation, as one part of a genome analysis process which is more complex than the basic string alignment: users define the preprocessing of the application data (protein strings), a specific alignment procedure, and the postprocessing of the output data (rating matrices) for searching in large databases for non-trivial genome similarities, i.e., string permutations which may in-

J. Dünnweber, S. Gorlatch, *Higher-Order Components for Grid Programming*,
DOI 10.1007/978-3-642-00841-2_4, © Springer-Verlag Berlin Heidelberg 2009

clude non-linear rearrangements. Genome analysis in bioinformatics represents another demanding class of real-world applications, where Higher-Order Components can significantly reduce the developments costs. An extensible genome processing HOC, called Alignment-HOC is presented in this chapter and experimentally evaluated on the grid. During the experiments with the Alignment-HOC, genome similarities in the ProDom database were detected that were not known previously. The ProDom database is a standard source for biological research and it contains about 70 MB of genome data. The knowledge about the newly detected similarities helps to better understand protein evolution and the functionality of protein sequences.

Both applications have high demands for computing power due to the data masses they process and are, thus, typical candidates for distributing computations in the grid. In both applications, we focus on programming methodology.

This chapter starts with the introduction of *Clayworks* – the first large-scale application case study that addresses the challenge of integrating the different computation and communication requirements of tightly coupled collaborative work and loosely coupled HPC applications – in Section 4.1. The different communication technologies that Clayworks employs for connecting between a central server, the clients and the grid are addressed, as well as the challenges that arise when these technologies are integrated with each other. The distributed 3-tier architecture of Clayworks, is described and the Deformation-HOC and its parallel implementation on top of the Globus middleware are explained.

The Alignment-HOC – a HOC for bioinformatics – is discussed in Section 4.2. The section introduces specific genome similarities, called *circular permutations* (CPs), which can be detected using the Alignment-HOC. It is shown how users can modify the Alignment-HOC for performing, instead of a general genome analysis, a search for circular permuted proteins, including experiments with real biological databases.

Conclusions from using HOCs in large-scale applications are drawn in Section 4.3

4.1 Clayworks: A Collaborative Simulation Environment

With emerging grid computing technologies, it has become a broadly used concept to distribute computations over the Internet in a variety of applications for business, science and engineering, which are often summarized under the term *eScience*. According to John Taylor's definition [cTay02], eScience "is about global collaboration in key areas of science, and the next generation of infrastructure that will enable it." Two important approaches in this area are: (a) *Computer-Supported Collaborative Work* (CSCW) environments which allow specialists to work together on a single project from different locations, and (b) *High-Performance Computing* (HPC) [bM+06].

The Clayworks system is designed as a distributed worksuite allowing to collaboratively model clay objects and execute deformation experiments with them.

In the modeling mode, several users concurrently model objects in a shared design workspace using virtual clay. Changes to objects are immediately shown at all user clients, which requires soft real-time communication and computation in order to provide a high level of responsiveness. In the simulation mode, clay objects are deformed when they crash into each other, which is simulated using a remotely located high-performance server.

In the CSCW part of Clayworks, several users, each connected via a graphical client to a shared workspace, model objects made of virtual clay:

- Users can create, delete, merge or modify the shape of objects, which will be immediately visualized at all connected clients.
- Objects can be grouped or locked for exclusive access, such that users can easily split up the work on complex objects among themselves and collaboratively create large scenarios involving a lot of objects.
- Objects can be saved to a database and reloaded later in other workspaces and projects, which allows to build a library of reusable objects.

In the simulation part, the computation of the deformations of moving clay objects is started on a multiprocessor remote server:

- Users assign a velocity and a direction vector to each object and define other simulation parameters.
- For each object, it can be defined whether the object is solid or should be deformable.
- Users can view the result of the simulation as a movie and freely move the camera and scale and rotate the scene.

Clayworks is a result of combining techniques from two active research areas: collaborative environments and HPC. Distributed collaborative applications allow experts from different locations to work together on a single project, while HPC, and especially the recent grid computing research, deals with accessing remote computational resources like processing power or storage space in a transparent way.

Deformation simulations among the moving objects can be executed, using the deformation algorithm described in [bDC04] as a basis. However, since the computation requires a lot of memory and processing time for larger scenarios, the algorithm was parallelized and embedded into a HOC. This setup enables the use of a server with a multiprocessor architecture for running the computations efficiently (see Section 4.1.2 for details).

This section gives an overview of the target application and the design of Clayworks. The 3-tier architecture (Client, CSCW-server, parallel computation) designed for the Clayworks implementation is introduced. The main operations which can be performed by the users are explained and the integrated workflow of cooperatively modeling clay objects and simulating their deformation in Clayworks is discussed.

4.1.1 The 3-tier Architecture of Clayworks

The main challenge in the development of Clayworks was the integration of the CSCW and the HPC part, which in some sense have contradictory requirements: The CSCW part is a soft real-time system which requires timely communication and computation for a high responsiveness of the application. However, the computations for the modeling are not very intensive and can be executed locally on a standard, modern desktop PC. In contrast, the simulation algorithm, which iteratively moves and deforms the clay objects, needs one or multiple HPC hosts with a very high computational power, but has no real-time requirements for the communication.

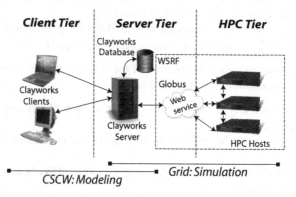

Figure 4.1 The 3-tier Architecture of Clayworks

In order to integrate these different requirements of the CSCW modeling and HPC in Clayworks, the 3-tier architecture shown in Fig. 4.1 was developed, allowing to build up each tier upon a different basis technology: the HPC tier is based on grid middleware, the server tier is based on RMI and JDBC [cSM06b], and the client tier is based on Java Swing. The server tier is in the middle, consisting of the Clayworks server and a database put in between the user clients and the HPC tier, thus decoupling the clients from the HPC host, i.e., users can store and reload 3D-objects without connecting to the HPC hosts. The use of a Web service for interconnecting the Clayworks server and the HPC host enables to flexibly exchange the HPC host according to the application requirements without affecting the client, as indicated by the dashed lines in the figure. This makes sense, when, e.g., instead of multiple compute nodes (e.g., of a cluster, as indicated in the figure) the parallel computations are outsourced to a single grid server, such as the SMP machine used in the experiments in Section 4.1.2.

The clients together with the server form the modeling part which is optimized for immediate communication, while HPC host(s) constitute the high-performance computing facility of Clayworks for the deformation simulations.

In the following, the three tiers and their respective functionality are briefly discussed:

Client Tier

The clients run at the users' desktop computers and provide an integrated access to all functionality of Clayworks. Users can connect to a shared workspace on a Clayworks server, model objects and observe, as well as chat about the work of other users in real-time. The Clayworks client makes use of OpenGL via the Java 3D API

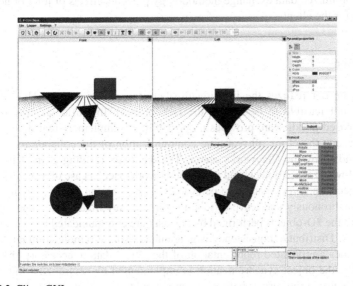

Figure 4.2 Client GUI

for the visualization of the workspace. It is important to notice that the clients are nothing more than visual terminals (shown in Fig. 4.2) which never perform any actions affecting data in the shared workspace locally.

The connection between the clients and the HOC is established via a Web service that is supplied with data and code parameters by sending it the corresponding identifiers. In the Clayworks application, a HOC is used for computing the simulation sequences; the identifiers which users send to it refer either to some material-specific transformation code (i.e., a standard, application-specific code parameter) or they refer to a 3D-object representation in a database.

Thus, the clients do not communicate the input data to the HOC directly in this application, but they communicate it to the Clayworks server, responsible for the synchronization of the data. The HOC independently downloads its input from the database.

The described decoupling between the Clients and HOC allows to maintain different representations of the workspace data (i.e., the coordinates and measures of the 3D-objects) at the Clayworks server and the HPC hosts running the HOC in the grid. Section 4.1.2 describes the voxel-based representation maintained inside the HOC used by Clayworks. The representation on the Clayworks server is rather hierarchically than graphically oriented. Instead of a polygonal mesh, only the types of the 3D-objects (sphere, cone, etc.) and their connections are stored using a tree structure (the *octree* [cBl06] which is quite common in computer graphics), optimally for the communication of modification commands in the network: When an object is affected by a change triggered by a client, not all the coordinate data, but only the modification command needs to be exchanged over the network. The HOC and the HPC servers in the grid are not involved in such client-side interactions at all. This optimized data exchange methodology prevents effects of jitter or network latencies for the clients.

Server Tier

This middle tier realizes the functionality between clients and the HPC host(s), which is not feasible to be run on the HPC host(s). In particular, current Web services-based middleware provides no possibility to implement real-time interactions of users; therefore, all real-time communication of the collaborative modeling is handled by the Clayworks server. The database, residing at the Clayworks server, allows to reuse objects and to recover a workspace in case of a server failure. Since objects and workspaces are persistently stored on the server side, users can work asynchronously and leave or join a session at any time. Once connected, all changes to the 3D-objects performed by any user are communicated to all users synchronously. Distributed consistency is guaranteed using a remote command queue which makes a replicated representation of the object data unnecessary [bM+06]. Principally, the object synchronization is implemented by centralizing the control over all objects at the server: before any command takes effect on the client side, it is communicated via RMI to the Clayworks server which locks all objects until the most recent changes on them have been committed and communicated back to all clients, similar to a 2-phase commit in a distributed database system [aGR93].

For simulations on the HPC host, the server prepares datasets for the remote computation; it starts and monitors the progress of the computations performed by the remote HPC host and finally returns the result to the clients.

HPC Tier

The HPC tier of Clayworks runs a parallel implementation of the clay deformation algorithm [bDC04] used for the simulation. This is realized using a HOC called Deformation-HOC: the Deformation-HOC allows its user to specify only material characteristics while it shields the user from the parallel algorithm (and its realization as a grid component) used to compute a simulation. The result is transferred back to the clients as a sequence of single images (see Fig. 4.3 for an example).

Users pick a default material definition offered by the Deformation-HOC implementation (e.g., the one for clay or solid objects) or define a new material, by specifying a material-specific density shift operation, which is a sequential step in the algorithm.

Figure 4.3 Screenshots of a Simulation Sequence

When users request the computation of a simulation, the Clayworks server commits all relevant data to the database (object coordinates and object measures, and additional simulation information, like the objects' velocities and movement direction vectors) and returns the corresponding database keys. These keys are sent to the Web service which connects the clients to the HOC which is used for computing the simulation in the grid. The deformation algorithm then iteratively moves and deforms the objects, resulting in a sequence of single simulation "screens". Upon completion of the simulation, i.e., when all objects have spent their kinetic energy for movement and deformation, the server downloads the simulation results from the Web service and sends them to the clients. To reduce the size of the transferred data, the ZLIB-compression utilities from the `java.util.zip` package are employed. Due to the low number of differences between subsequent pictures in an animation, the size of the result files, which must be transferred over the network, can typically be less than a hundredth of their original size in average, enabling a quick transfer between the Clayworks database and the HOC.

4.1.2 The Deformation-HOC for Parallel Simulations

The simulation algorithm employed by Clayworks can be used for the parallel implementation of a broad class of spacial simulations. It is not only suitable for crash tests but for computing realistic presentations of any scene wherein an arbitrary material is deformed, e.g., the diffusion of gases or the freezing of fluids. The algorithm exhibits a recurrent control flow structure. Therefore, it is a typical candidate for a reusable implementation as a HOC. This section presents the concept and implementation of the Deformation HOC, which was developed for this purpose.

The algorithm implemented by the Deformation-HOC in Fig. 4.4 represents an enhanced version of the algorithm from [bDC04]. The HOC runs the algorithm in parallel on multiple partitions of the object data and, in contrast to [bDC04], it does not require the presence of a solid tool object, i.e., all objects in the virtual universe can be built of clay.

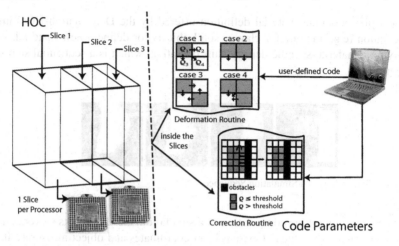

Figure 4.4 Computation Schema of the Deformation-HOC

Data Representation in the Deformation-HOC

Specifically for the Deformation-HOC, there is, besides the octree-based *polygonal* representation which the users collaboratively work on, another *voxel-based* data representation in a three-dimensional grid. Each voxel represents a cell in the simulation space as a data record holding its coordinates and attributes such as the material density, etc.

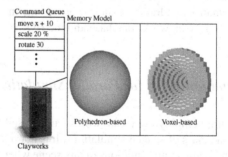

Figure 4.5 Two Object Representations on the Clayworks Server

Figure 4.5 shows a sphere-shaped object in the two different formats: the polygonal version is shown left and the coarser presentation on the right illustrates the voxel-based version. In the cut-away view of the voxel-based version, it can be seen that only the visible external part of the object is covered by voxels while its inside is unfilled, thus reducing the data size of this format. The polygonal representation is used in the modeling part and allows to build the objects in an intuitive way by defining vertices and faces. The voxel-based representation is required by the parallel simulation algorithm in the Deformation-HOC.

Upon the start of the simulation, the Clayworks server converts the polygonal objects into the discrete voxel-based representation using the algorithm described in [bH+98]. This initialization step forms only a very small fraction of the simulation computation. It is therefore performed directly on Clayworks server. This way, only a single representation of the object data is required on the remote HPC host(s) in the grid which perform(s) the largest part of the computations in parallel. Upon completion of the parallel computations, the Clayworks server runs the *Marching Cubes* algorithm [bLC87] for re-polygonalizing the objects on each scene of the simulation sequence.

The Multithreaded Deformation-HOC Implementation

In the following description of the Deformation-HOC, it is assumed that the component is implemented as a multithreaded Java program. This way, the Deformation-HOC was implemented for the experiments described below and a single SMP-server that maps Java threads to different processors was used as the HPC hosting platform. However, many alternative implementations of this component are possible (e.g., for a cluster or a network of workstations) and can be connected to Clayworks using the same Web service.

In the presented version, the Deformation-HOC can also be used to compute multiple independent simulations on multiple distributed grid servers simultaneously, but each single simulation is then computed using one dedicated server.

The Internal Multilayer Process

Internally, the Deformation-HOC performs parallel computations which are invisible to the users: the voxel-based representation of the objects is held in a 3-dimensional array, the *cubic voxel universe*. Each element of this array is an instance of the Voxel-class, holding the two attributes density and velocity. For each voxel, the density attribute stores the density of the clay at the voxel. All voxels have the same constant volume, which depends on the granularity of the cubic voxel universe, which must be specified before the simulation starts. Before the start and after the completion of the algorithm, the clay density of all voxels ranges between 0 and 1, but during the computation, values above 1 are possible. Voxels with a density value above 1 are called *congested*. The impact of congested cells on the computation is discussed below. The second attribute of the Voxel-class, velocity, is a vector describing the movements taking place in the universe. These movements are always linear shifts, triggered by the Clayworks users.

In the Deformation-HOC, the clay deformation algorithm [bDC04] is extended by simulating energy reduction, which terminates the deformation process in a natural way without requiring the user to stop it.

The Deformation-HOC performs a 3-layer traversal of the simulation space (i.e., three passes). In each single traversal, all voxels are visited once and new density values are computed for them. During the first layer traversal, the shifting of voxels is computed. As long as the position of a voxel after a shift does not collide with the position of another voxel, there is no clash and the direction of the velocity vector belonging to this voxel is preserved. The magnitude of the vector is altered by multiplying it with an energy reduction factor α in each traversal. To find a good assignment for the energy reduction α, multiple different factors were tested, including simple linear ones. Empirically, it was verified that the reduction of energy during a shift is simulated very realistically when the exponential function is used to compute the energy reduction factor α, as follows:

$$\alpha := exp\left(\frac{-1.0 * \rho_1}{\rho_2}\right) \tag{4.1}$$

where ρ_1 and ρ_2 are the density values of the clashing voxels. When voxels clash into each other, a new velocity vector is computed using the formula

$$\delta := \frac{1-\alpha}{2}\delta_1 + \frac{1+\alpha}{2}\delta_2 \tag{4.2}$$

where δ_1 and δ_2 are the original velocity vectors of the clashing voxels. α adheres to definition (4.1) reflecting the intensity of the clash, which is reduced proportionally to the reduction of energy.

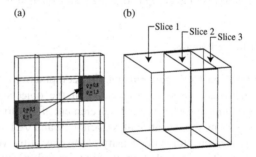

Figure 4.6 Density Shift and Sliced Partitioning

The computations in this layer are performed as a simple vector addition, allowing for a density congestion in the resulting voxels (see Fig. 4.1.2, where a shift between two voxels results in a density above the example threshold 1). The purpose of the successive layers is to adapt the results of the first layer, such that congested voxels distribute their density overplus among their neighbors.

Each layer of the deformation algorithm is parallel in nature, as the cubic voxel universe can be partitioned into sub-cuboids (slices) as shown in Fig. 4.1.2, and each layer can be applied in parallel to the slices, while only the ordering of the layers must be preserved. The partitioning depends on the distribution of the clay objects specified by the user. Synchronization and communication is required whenever an operation inside one slice affects voxels in another slice. The irregularity of this partitioning is due to the load-balancing: smaller slices (slices 2 and 3 in the figure) contain more congested voxels, such that the number of dependences between slices is approximately balanced.

Code Parameters of the Deformation-HOC

In the *Deformation-HOC*, only the procedures during density shifts in the first layer and the procedures to correct overfull voxels in the remaining layers are application-specific. Therefore, the *Deformation-HOC* takes one customizing parameter which carries a serialized Java class.

```
1:  public interface DeformationParameter {
2:      public boolean isSolid(int objectID);
3:      public double getMaxDensity( );
4:      public double powTrans(int distance, int power);
5:      public double densTrans(int layerNr, double dens); }
```

Figure 4.7 Java Interface for a Deformation-HOC Code Parameter

This class must be accessible using the interface shown in Fig. 4.7. The methods in this interface are as follows:

isSolid (line 2): returns, for a given objectID, whether the respective object is built of a solid material and therefore not affected by the deformation process;

getMaxDensity (line 3): determines when a voxel is overfull. When a value above 1 is specified, less corrections must be performed, speeding up the higher-layer computations, but the accuracy of the simulation is reduced this way;

powTrans (line 4): converts kinetic energy into deformation energy, as it happens during the deformation process;

densTrans (line 5): is used to distribute density among neighboring voxels in layers > 1, i.e., within a correction.

Figure 4.8 presents a closer view at the parameter code (familiar from Fig. 4.4). The DeformationParameter, used to customize the Deformation-HOC, is composed of a deformation routine (Fig. 4.8, left) and a correction routine (Fig. 4.8, right). The Deformation-HOC always tries to move mass out of a congested voxel, as shown for case 1 (in Fig. 4.8, left) first, and, if not possible (depending on the density values of the neighboring voxels) choosing, as an alternative, case 2, case 3 or case 4 (in this order), since, this way, mass movements result in as little new congested voxels as possible. The user-defined powTrans conversion is called to compute the density updates for each voxel. Figure 4.8, right, shows how a single correction layer works, i.e., the process which is repeated until no congested voxels remain.

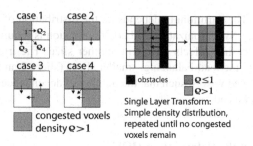

Figure 4.8 Modifying the Density Shift Inside the Deformation-HOC

If no `DeformationParameter` parameter is specified by the user, the Deformation-HOC uses the `DefaultDeformation` class, wherein, e.g., the `powerTrans`-method is defined using the exponential function, as shown in lines 2–3 of Fig. 4.9.

```
1: double powTrans(int distance, int collisonPower)  {
2:   return Math.exp( (-1.0 / (collisionPower * 2) )
3:            * (distance * 2);  }
```

Figure 4.9 Default powTrans-method for Clay Deformation

Also the other methods from the above interface are defined following the definitions (4.1) and (4.2) in Section 4.1.2; i.e., as long as no different `DeformationParameter` is specified, the *Deformation-HOC* computes the deformations for objects of a clay-like material.

Writing a specific `DeformationParameter` for a different application requires some knowledge about the physical properties of the material which is simulated. When, e.g., a steel brick is simulated, the required `DeformationParameter` is derived from the `DefaultDeformation`-parameter and the methods `isSolid` and `powTrans` are overridden, such that `isSolid` returns `true` and `powTrans` and `densTrans` methods return constants, depending on the weight of the brick.

The *Deformation-HOC* facilitates code reuse and customization. Its full implementation comprises approximately 600 lines of Java code plus 300 lines of XML code for the WSRF support. When a new `DeformationParameter` is derived from the `DefaultDeformation`-parameter for different simulations, a programmer using the *Deformation-HOC* has to write no more than 20–30 lines of Java code.

Experiments with the Deformation-HOC

Some experiments were conducted using a SunFire 880 SMP server with 8 UltraSparc-III processors running at 1200 MHz.

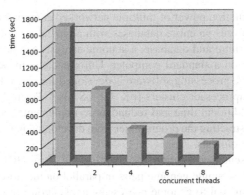

Figure 4.10 Experimental Results

Figure 4.10 shows the average runtime for computing a deformation of two clay cubes (see Fig. 4.3). As can be seen, there is an almost linear speedup for up to 4 concurrent thread. For a higher number of concurrent threads similar results can be expected when also the problem size (i.e., the number of object in the simulation space) is increased. For 8 threads, e.g., a speedup of 7.2 was measured, for a "good" case, i.e., for a scene wherein the objects were evenly distributed.

In the Clayworks example the HOC processes data which is produced by multiple users who concurrently manipulate objects. The following section presents an example where the input data is taken from different databases, selected by the users. Since the databases can be quite large, the data masses the HOC has to cope with are considerably higher than in the Clayworks application.

4.2 Protein Sequence Analysis with HOCs

This section deals with another real-world grid application, the analysis of large amounts of genome data, and introduces, for this purpose, the Alignment-HOC. The Alignment-HOC offers the distributed implementation of a generic alignment algorithm and allows users to plug in their own code, which makes it easy to study genome similarities in different databases or to run a specific genome processing algorithm on a biological database (e.g., 3D structure prediction).

4.2.1 The Alignment Problem in Bioinformatics

Genome processing algorithms which are used for sequence alignment or protein structure prediction typically compute one or more result matrices and have the time complexity of $\mathcal{O}(n^2)$ or higher for sequence length n [bNW70, bSW81]. Working on a large amount of data sequentially quickly causes an unreasonably long runtime.

Therefore, the calculation power of multiple networked computers is indispensable to run pairwise analyzes on entire databases with a typical size of almost 100 MB. Depending on algorithm and database, a genome analysis can take several months of calculation time on a standard computer. Unfortunately, most of the available software for Genome processing software is developed for single PCs [bW⁺05] (or homogeneous multiprocessor clusters [bK⁺02, bS⁺04]). Porting such software to the grid requires additional time and redundant re-implementations of the same or similar software, distracting the programmer from developing new algorithms which are interesting for biologists.

Section 2.4 presented a solution for adapting the Farm-HOC to compute sequence alignments for single genome pairs in parallel on the grid. While sequence alignment is fundamental to genome processing applications, most genome analyses are more complex and require, besides the processing of a multitude of genome sequence pairs, e.g., a preprocessing of the sequences and a postprocessing of the output. This section presents a new HOC, specifically developed for genome processing. While the user is free to arrange custom pre- and postprocessing operations of the input, this HOC always computes an alignment and is, therefore, called Alignment-HOC. The purpose of the code parameters of this HOC is to customize the Alignment-HOC for different kinds of similarity detections. This is demonstrated in the following by a case study of searching so-called *circular permutations* in genome databases. Using an appropriate post-processing parameter (called *traceback*), the Alignment-HOC is also optimized for improving its reliability.

4.2.2 Circular Permutations of DNA

Circular permuted protein sequences (CPs) occur in a number of protein families [bW⁺05] and can be found in all large databases of protein data. Their linear order may be quite different, but the 3-dimensional structure of their resulting protein and its biological functionality are often the same.

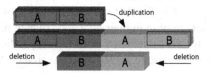

Figure 4.11 Possible Development of a Circular Permutation

In Fig. 4.11, *A* and *B* are arbitrary subsequences of a protein sequence. The figure shows an example of one possible development of circular permutations, by doubling the original sequence and inserting afterwards new start and end codons (tri-nucleotide codes that define the begin and end of a gene expression [bJe99]). Such a circular permuted sequence consists of two parts from the original sequence, but in a different order. Another way of CP development is assembling two proteins from existing sequence fragments [bBu02].

The problem with circular permuted sequences is caused by the non-linear rearrangement of the amino acid order. Different from other mutations like insertions or deletions of single amino acids in the linear sequence string, a CP shifts the beginning of the sequence to the end, thus, radically changing the original linear order of the sequence. Standard alignment algorithms will not detect a significant similarity between the original and the circular permuted sequence (shown in Fig. 4.12), although the tertiary structures (the 3-dimensional folding of the amino acid chain) of the resulting proteins may be nearly the same.

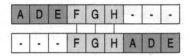

Figure 4.12 Standard Alignment of a Circular Permuted Sequence

Similarities between circular mutated sequences and their related sequences are usually not marked in sequence databases, making it difficult for a scientist to work with such sequences. In order to find these similarities, the whole database must be processed by a specialized algorithm adjusted for the non-linear sequence rearrangement of CPs.

4.2.3 The Alignment-HOC and its Code Parameters

The Alignment-HOC (see Fig. 4.13) was developed for genome analyses like, e.g., the detection of circular permutations (as explained above). Actually, the Alignment-HOC can be used for detecting CPs (i.e., sequences which are highly similar after the CP has been made undone) for different kinds of input: either for sequences of protein domains (domains are functional units in a protein consisting of up to several hundred amino acids) or for the underlying amino acid sequences. Because domains consist of many amino acids, the domain sequences of a protein are significantly shorter than its amino acid string. This reduces the complexity of calculations performed on domain data but also reduces the amount of protein information available for the similarity detection.

Both CP detection algorithms have their advantages and disadvantages in different applications. Therefore, the user can vary between both versions via the code parameters of the Alignment-HOC. Switching between both versions (via a Web service, as explained in Chapter 3) only affects the operations that differ when either domains or amino acid strings are being processed, while the main component code is reused in any application.

The Alignment-HOC has three code parameters.

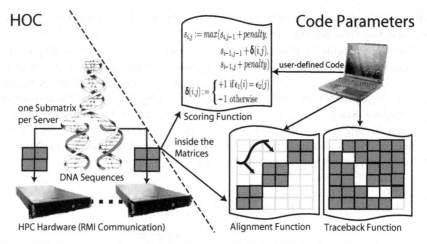

Figure 4.13 Computation Schema of the Alignment-HOC for DNA Similarity Detection

Parameter 1: Scoring Function

Via this parameter users can modify the scoring function, i.e., the operations used to compute a single element in a scoring matrix using the same `Binder`-interface as shown in Fig. 2.31 (Section 2.4.2). In this matrix, each element rates differences between pairs of protein residues (i.e., elements of the genome code) and is applied to the two subsequences, delimited by the matrix elements' indices. In most cases, the scoring function complies with the default definition for a Needleman Wunsch Alignment [bNW70] (shown in Section 2.4.2). However, the Alignment-HOC is more flexible than the simple Wavefront-HOC (i.e., the adapted Farm-HOC from Chapter 2). The Alignment-HOC does not only allow its users to vary gap penalties [bH+90] but also a different scoring schema can be implemented, e.g., according to the lengths and frequencies of collective insertions and deletions. Such a scoring schema can lead to a more accurate alignment, since the evolution of genomes involves more complex processes than point mutations (i.e., a change of a single base into another base [bCa07]).

Parameter 2: Alignment Function

This parameter allows the user to control the course of actions in a genome processing application using the Alignment-HOC. For this purpose, users may implement the interface shown in Fig. 4.14. If a standard Needleman Wunsch Alignment should

```
1:  public interface Aligner extends Serializable {
2:  public char[] preprocess(char[] sequence);
3:  public char[][] align( char[] seq1, char[] seq2,
4:                    boolean postprocess );  }
```

Figure 4.14 Aliger-Interface for the Alignment Function

be computed, there is no need to care about these methods, but the Alignment-HOC will run the default implementation (based on the JAligner library [cAM07]). In a custom alignment, users may convert the input sequences in the `preprocess`-method (line 2), e.g., for filtering out information that is not relevant in an application or an on-the-fly translation from DNA to amino acids or vice versa. The `align`-method allows users to adapt the computation of the scoring matrix, e.g., for optimizing it for a particular grid platform type: as shown by experiments in Chapter 2, computing a single scoring matrix using the wavefront pattern scales well on shared-memory servers, while computing multiple independent matrices in parallel is preferable in a distributed-memory network. The `postprocess`-flag (line 4) is used to enable/disable a postprocessing, defined via the traceback function which is supplied to the HOC as its third code parameter.

```
1: public interface Traceback extends Serializable {
2:    public int[] traceback( char[] directionMatrix,
3:                              int width, int height);  }
```

Figure 4.15 Traceback-Interface for the Traceback Function

Parameter 3: Traceback Function

Via this parameter, the user can specify how a traceback path is detected, i.e., a path following high values in the scoring matrix which starts at the bottom/right side of the matrix. Traceback paths are encoded as flat arrays of coordinates, as obvious from the interface shown in Fig. 4.15. The default traceback results in one main path that always starts in the most bottom right element (the *global* score which rates the similarity of the sequences over their total length) and runs in a more or less diagonal manner through the whole scoring matrix. Typically, there are variations in the diagonal path, since the compared sequences have point mutations which are reflected by an uneven distribution of high scores in the matrix. Figure 4.16 (left) shows a traceback running straight along the middle diagonal, indicating that the compared sequences consist of the two equal chains *A* and *B*. Via the traceback function, Alignment-HOC users can modify the processing of the scoring matrix resulting from a sequence alignment to perform different kinds of genome similarity detections.

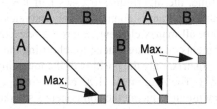

Figure 4.16 Standard Alignment (*left*) and CP Detection Algorithm (*right*)

4.2.4 Using an Alternative Traceback

The right part of Fig. 4.16 shows what the Alignment-HOC does in the traceback
step when we specify the traceback code parameter for CP detection (as described in
Section 4.2.2) instead of the default traceback. The CP code parameter is included in
the Code Service of the HOC-SA, i.e., users can run this kind of genome processing
by simply selecting this alternative traceback parameter and are not required to write
new code for this application.

In case of a circular permutation between two compared sequences, the first part
of each analyzed sequence is strongly similar to the last part of the other sequence
(or the parts are even equal, as in the depicted example). Instead of one main traceback path, an alignment between two circular permuted sequences has two traceback
paths starting from two *local* maximum scores, one in the bottommost row and the
other one in the rightmost column of the scoring matrix (marked in the figure with
'Max'). If such two maxima cannot be located, since there is almost no variation in
the values on the outermost matrix row and column, the sequences are not circular
permuted and no further testing is necessary.

When the traceback paths run (almost) along shifted diagonals of the matrix, as
in the figure, a CP is detected. The traceback code parameter for detecting CPs tests
for the matrix whether the criterion *traceback paths run along shifted diagonals* is
fulfilled as follows: since shifted diagonals pass three quadrants of the matrix and
never intersect each other, it starts from the two maxima, follows the high scores and
counts the number of quadrants passed on these paths. The test is positive if there is
no intersection (i.e., a common element) and both paths pass three quadrants.

4.2.5 Optimizations of the Alignment-HOC

As compared to the many other implementations of sequence alignment (including
the Wavefront-HOC in Chapter 2), the Alignment-HOC offers some optimizations
which were specifically developed to increase the efficiency of CP detection using
the Alignment-HOC. The following paragraphs show how these optimizations work
and how users can benefit from them by selecting the corresponding code in the
Code Service, without dealing with the implementation themselves.

The Alignment-HOC stores in the main memory only the part of the alignment
data that is necessary for CP detection. Instead of working on both doubled sequences at the same time, the algorithm processes half the sequences separately
and, thus, when a matrix quadrant is computed, only the relevant half of the input
is loaded. Moreover, not all matrix element values are relevant in the application,
but only the matrix elements which are crucial for choosing the traceback direction,
i.e., for each row of the matrix, only the position of the maximum element is stored,
as suggested in [bGo82]. The Alignment-HOC allows to apply these application-
specific optimizations without reducing the component's standard functionality. All
optimized code is placed inside the code parameter for the alignment function (parameter 2, see Section 4.2.3) which users can replace by the default step on demand.

Doubled Input for Increased Sensitivity

To avoid false-positive detections of CPs, both compared genome sequences should be doubled. before the scoring matrix is calculated. After the duplication the trace-back results in extended paths.

Figure 4.17 shows how this path extension is used to increase the sensitivity of CP detection: the lengths of the four line segments which the traceback paths mark by their intersections with the inner borders of the matrix quadrants are compared. Only if the corresponding lengths (i.e., the segments of a single line, such as α and β on the inner vertical border) have nearly the same ratio, the CP test is positive. Doubling the sequences obviously results in a matrix that is four times larger, but experiments justify the higher computation costs: tests without doubling the input [bW$^+$05] delivered too many false-positive results.

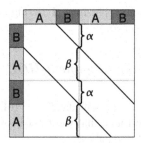

Figure 4.17 Increased Detection Sensitivity

Noise Reduction for Increased Accuracy

The Alignment-HOC can process domains and also amino acid sequences. Because of the high number of different domains that occur in a domain sequence (there are several thousands of known domain strings in biological databases), a match, i.e., the occurrence of equal input elements, in the scoring matrix is relatively rare when domain data is aligned. In the case of aligning amino acid sequences, the probability of a match is 1:20, since only 20 different amino acids exist. A high number of matches in the scoring matrix produces noise, i.e., variations in the distribution of high scores, which makes the detection of the two characteristic shifted diagonals that indicate a CP quite difficult.

Figure 4.18 shows the optimization in the Alignment-HOC that helps to avoid the described similarity noise by rating the matches. Instead of noise suppression in the result matrix, the Alignment-HOC reduces noise while calculating the scoring matrix. Whenever input elements match, a different value than the scoring function output is assigned to the corresponding matrix element, e.g., score 1 for a simple match, score 10 for a double match (i.e., matching elements plus a match in the upper left neighbour cell), score 20 for a triple match and score 50 for four

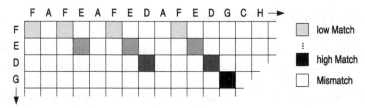

Figure 4.18 Reducing Noise by Using Dynamic Match Values

successive matches on the main diagonal (i.e., top left to bottom right). Thus, the rating counts neighboring matches for rating the degree of a match. The exact values chosen for the rating schema are arbitrary, except for the condition that they range approximately within the scoring function's codomain to effectively suppress low similarities in the aligned sequences. This simple noise reduction schema has been experimentally proven to satisfy the demands of CP detection, while other genome similarity analyses require more sophisticated filering methods [bH+02] including, e.g., wavelet transform (Chapter 3).

4.2.6 Experiments with the Alignment-HOC

Using the Alignment-HOC, the protein database ProDom was scanned [bL+07] and genome similarities were found that were not known previously, since earlier projects never processed the full database.

ProDom Version	Raspodom	Alignment-HOC
2003.1	36	129(40)
2004.1	93	850(192)

Table 4.1 Number of Detected CPs in Different Databases

Table 4.1 shows the number of CPs detected by the Alignment-HOC and, in turn, by a program from a related research project (called *Raspodom* [cW+03]) for two versions of the ProDom database. The numbers in parentheses are those CPs where two ore more protein domains are involved in the sequence rearrangement. In these sequences, the detected CPs span at least two functional parts of the protein data and, thus, deliver strong evidence that the sequences are closely related.

Two of the newly detected CPs are shown as dot plots in Fig. 4.19. In each dot plot, the two characteristic shifted diagonals (see Section 4.2.3) are observable (compare Fig. 4.16). The new CP, shown in the dot plot on the left, compares the sequences *'Peptide synthetase MBTF'* (O05819) and *'FxbB'* (O85019). Both are parts of mycobacteria genomes, but only the function of *'FxbB'* has been known in the Swiss-Prot database as part of an adenosine monophosphate (AMP) binding enzyme family. The relationship between *'Peptide synthetase MBTF'* and *'FxbB'* of being circular permutations of each other leads to the hypobook that *'Peptide synthetase MBTF'* is also part of the same AMP binding enzyme family with a similar function.

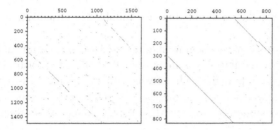

Figure 4.19 Dot Plots of the Newly Detected Circular Permuted Sequences

The dot plot in the right part of the figure compares the two nameless sequences Q69616 and Q9YPV1. Both are listed in the Swiss-Prot database as polymerases, but only Q9YPV1 is a known DNA-polymerase. The newly detected circular permutation between the two sequences indicates that Q9YPV1 is also a polymerase interacting with DNA.

4.3 Conclusions from Using HOCs in Large-Scale applications

In this chapter, it was shown by two practical examples, how HOCs help programmers porting real-world applications to the grid. The first example, Clayworks, is a CSCW application that combines an interactive multi-user environment with high-performance computing. The employed HOC, the Deformation-HOC, is used to outsource the compute-intensive simulation process to the grid. The second example is the genome sequence processing via the Alignment-HOC.

The Alignment-HOC is able to handle the pairwise processing of hundreds of megabytes of data (as present in total genome databases) by distributing the computations. The calculation power offered by the Alignment-HOC makes it possible to keep up with the exponentially growing amounts of data in genome sequence databases used in biological data analysis applications. The development of new problem-specific code parameters can be easily done because only simple Java interfaces have to be implemented. The data distribution is transparent to the user.

The main contribution of the Clayworks worksuite is that it tightly integrates collaborative modeling with a grid component: The Deformation-HOC which computes the deformation simulations in parallel. Thus, end users can collaborate and easily access high-performance grid servers in a transparent way.

When compared to other distributed problem solving environments like CU-MULVS [bW+06] or NetSolve/GridSolve [bS+05], Clayworks' novel feature is the support for tightly-coupled, synchronous collaboration with soft real-time deadlines. The 3-tier architecture of Clayworks satisfies the different requirements of collaborative modeling and of HPC. The real-time requirements are combined with the HPC infrastructure in a transparent way for the end users. Furthermore, the transformation of the clay objects from polygonal to voxel-based representation and vice versa allows to use the representation best suitable for visualization and computation, respectively.

The concept of seamless integration of remote HPC servers into application software can be expanded into novel areas beyond traditional industrial and scientific applications. In order to make such software suitable for the mass market and non-experts in the area of computing, the access to HPC resources has to be made as transparent as possible. As shown in this chapter, HOCs are a promising step into this direction.

Chapter 5
HOCs with Embedded Scheduling and Loop Parallelization

This chapter deals with two advanced topics of component-based grid programming: (1) automatic scheduling (i.e., the mapping of application tasks to processing nodes), and (2) automatic parallelization of code. Both types of automatism are features that increase the degree of abstraction of a component technology and lead to user-transparency, i.e., application programmers are less involved in technical details of the grid platform.

In the HOC applications shown so far, it was the application programmer who selects target nodes in the grid (by calling `configureGrid(nodeList, ...)`, as, e.g., in the example application shown in Section 2.1.4). If dependences in an application require an adaptation of a component for the parallel execution of that application (as, e.g., in the alignment example in Section 2.4), it was, so far, also the application programmer who must specify this adaptation.

In this chapter, two tools are introduced that can free the HOC user from these tasks: (1) the KOALA scheduler [bME05] that can select proper execution nodes and (re-)schedule application tasks, according to the requirements of the HOCs employed in a grid application, and (2) the LooPo loop parallelizer [cUP97] that can detect data and control dependences in the code parameters sent to a HOC, allowing the user to execute any nest of sequential loops in a code parameter in parallel without any manual adaptation.

Both tools, KOALA and LooPo existed before the first HOC experiments on the grid were conducted. Both tools were developed with the aim of simplifying the development of parallel and distributed applications and offer a functionality that is orthogonal to the functionality offered by HOCs. This chapter shows, where KOALA and LooPo are useful for the HOC developer and what their impact on the performance of HOC applications is.

Section 5.1 introduces the KOALA scheduling infrastructure including the utilities, which KOALA employs for co-allocation (i.e., the allocation of grid nodes located on multiple different sites), as well as enhancements of KOALA for the efficient, user-transparent scheduling of HOC applications. A special bandwidth-aware algorithm for HOC application scheduling based on cost-functions is also explained in this section. Different cost functions for scheduling HOC applications

J. Dünnweber, S. Gorlatch, *Higher-Order Components for Grid Programming*,
DOI 10.1007/978-3-642-00841-2_5, © Springer-Verlag Berlin Heidelberg 2009

are compared and the consequences of scheduling HOCs in a platform that is not *space-shared* are analyzed, i.e., when grid nodes are potentially running multiple HOCs at the same time. As experimentally proven, *user-Service Level Agreements* (uSLAs) can prevent the overloading of nodes in such a scenario. Surprisingly, the experiments with running HOC applications in the enhanced KOALA infrastructure also prove that user-transparent scheduling does not only preserve the performance of applications but actually leads to performance improvements. Human users (or traditional tools) typically make the scheduling choices according to the data emergence in an application, such that the host storing most of the data is made responsible for most of the data processing. Due to the structured processing patterns of HOCs, the scheduling system can predict in what phases most of the data exchange takes place in a HOC application, which enables a smarter mapping. KOALA takes the dynamic communication behavior of the application into account (which is independent from the initial, static distribution of input), leading to a very efficient scheduling, where the user is fully freed from explicitly choosing execution nodes.

Section 5.2 introduces the LooPo-HOC [bL$^+$06], a new component that embeds LooPo [cUP97], a tool for the automatic parallelization of program loops. Besides LooPo itself, the LooPo-HOC contains the mwDependence service [bT$^+$08], also explained in this section, which provides an adapted version of the master/worker processing pattern (as used in the Farm-HOC in Chapter 2) for executing ordered task graphs, taking dependences into account. A parallel SOR solver for linear equation systems is used as a case study. Measurements prove that the automatically transformed code scales well on distributed grid nodes.

Section 5.3 is about the integration of HOCs into the ProActive/GCM and about the relation between the HOC-SA and alternative component technologies, like SOFA.

Section 5.4 concludes the chapter with a closing discussion on the perspectives of combining HOCs with distributed processing tools such as LooPo, KOALA or, e.g., an implementation of distributed shared memory (DSM), as it is provided by JavaSpaces [cSM07e].

5.1 User-Transparent Grid Scheduling

Scheduling application tasks in the grid (e.g., the code units processed by the workers of the Farm-HOC, introduced in Chapter 2) is a complicated problem. Indeed, in all the introductory application examples of Chapter 2, it was supposed that there is not a vast quantity of tasks and all tasks produce a similar workload, such that the scheduling is manageable by letting the user directly select the processing nodes. However, in large-scale grid applications, e.g., the bioinformatics example from Chapter 4, picking the processing nodes usually requires specific knowledge about the application and the data that is being processed [bBE01]. In particular, the volume and the frequency of communication are important for scheduling: for the efficient assignment of tasks to processing nodes, and, potentially, the replacement of nodes during application runtime (for load-balancing or failover reasons), the placement of tasks cannot be arbitrary, since tasks with high data emergence require grid nodes with a very fast network connection.

The communication behavior of an arbitrary application cannot be foreseen during its setup, and especially in large-scale applications, it is often the task of the application developer or the end user to provide information about the application's communication properties, which complicates grid application development. While components like HOCs simplify the development of applications, they also imply a change of focus for scheduling: components and their compositions, instead of arbitrary application tasks, are mapped to the available grid nodes.

This section presents a user-transparent scheduling technique for HOCs, i.e., a technique that frees the end user from choosing processing nodes. The presented approach to user-transparent scheduling makes use of:

1. scheduling cost-functions that take into account the communication costs,
2. a workflow-reuse technique for avoiding the need for a repeated scheduling phase when a workflow recurs, and
3. an aggregated submission technique for avoiding multiple submissions for HOCs/code parameters to an execution site.

The KOALA system [bME05] is used as the basic scheduler. KOALA supports co-allocation – the scheduling of parallel applications onto distributed grid sites. Besides the KOALA system itself, this chapter presents some enhancements to the KOALA scheduler for HOC applications, taking their specific communication properties into account.

5.1.1 The KOALA Grid Scheduling Infrastructure

The scheduling infrastructure used in this chapter (shown in Fig. 5.1) is composed of the KOALA scheduler plus remote schedulers (typically, the *Sun Grid Engine* [aC+04], which is used for scheduling work units on a single grid site) and uses the Globus Monitoring and Discovery Service (MDS [bH+05]) for workload monitoring.

KOALA has been designed and implemented by the PDS group in Delft in the context of the *Virtual Lab for e-Science* project (VL-e [cB+02]). The main feature of KOALA is its support for co-allocation, i.e., simultaneous allocation of resources in multiple clusters to a single application consisting of multiple stages. KOALA supports processor and memory co-allocation.

In the KOALA infrastructure shown in Fig. 5.1, every host used for submitting jobs has at least one KOALA Runner (a utility program for task submission) in operation, while one centralized grid host runs the KOALA Engine. All runners communicate with the engine via a proprietary light-weight protocol based on exchanging two distinct types of messages (called *requests* and *answers*). Applications are executed by submitting single-CPU tasks (e.g., a code parameter of a HOC, such as a

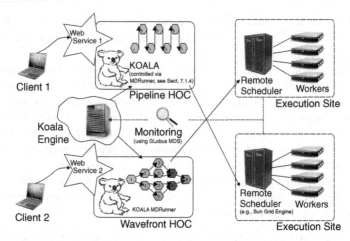

Figure 5.1 Overview of the KOALA Scheduling Infrastructure

pipeline stage) to the runners which handle the reservations of resources. The runners take as their input partially filled-in job descriptions in the Globus-typical *Resource Specification Language* (RSL) format, without the `resource manager contact` string specified [bC+98]. This RSL-document is passed to the KOALA engine which selects the appropriate execution sites, based on previously gathered monitoring information (collected using Globus MDS [bH+05]). The KOALA engine puts scheduling choices into effect by completing the RSL documents, i.e., by inserting the resource manager contact strings. The tasks are sent back to the runners, packaged together with the completed RSL descriptions (such a package is a *job*, in the Globus terminology [bC+98]). Finally, the execution sites trigger the execution of the jobs on the grid using the *Globus Resource Allocation Manager* (GRAM) [bC+98].

KOALA provides different kinds of runners, specialized for different types of jobs:

- The KRunner is the most basic KOALA runner, used for executing standard GRAM jobs which are programs that are executed by invoking them directly from the UNIX shell.
- The DRunner is used for starting multiple jobs that communicate via MPI and require an appropriate runtime environment (this runner is called **DRunner**, since it is based on Globus DUROC [cGl05]).
- The GRunner is used for scheduling Java jobs, i.e., Java programs which are executed within a virtual machine and make use of RMI (or the RMI-based Ibis library [bR+05]) for communication (this runner is called **GRunner**, since it internally uses the `globus-job-run` script [cTD07]).

Per default, KOALA makes use of the *Close-to-File* policy (CF) or the *Incremental Claiming* policy (IC) for scheduling jobs. Under CF, execution nodes are chosen depending on the estimated time of transferring the input files to the nodes, while under IC, the execution nodes are selected at runtime [bME05].

To integrate HOCs in the KOALA infrastructure, a new runner was created, called MDRunner (for *modified* DRunner). The MDRunner is an extended version of the DRunner which provides a new communication and input-file-aware scheduling policy.

5.1.2 Extensions of KOALA for User-Transparent Scheduling

Applications using HOCs have well-defined requirements in terms of data distribution and communication schemata, i.e., which data is sent when, and to what processors, is completely determined by the HOC type and does not depend on the particular application where that HOC is used. Many different applications can be built using the same HOCs or combinations of them. The user-transparent scheduling methodology for HOC-based applications presented in this chapter works independently of application-specific code and data parameters.

The following techniques are combined for the efficient, user-transparent scheduling of HOC applications:

- *The Cost-Functions* take in account the different characteristics of the grid platform where a HOC application is executed. Various costs functions are employed for achieving an adequate distribution of code units to the grid nodes.
- *The Reuse of Schedules for Similar Workflows*. Computing a schedule is expensive in the grid. In the presented approach, multiple subsequent applications that use the same HOCs are mapped onto the same resources, whenever possible (while a single application can run multiple instances of the same HOC in parallel, using different resources).
- *The Aggregated Submission of Code Units* exploits the fact that in many applications multiple code parameters carry the same code (see, e.g., Chapter 3, for an example where two application stages run alternately in the Lifting-HOC). This chapter shows how unnecessarily repeated submissions of code can be avoided.

When KOALA is used for scheduling HOCs, users benefit from an enhanced code transfer methodology. Both, the HOC-SA (introduced in Chapter 3) and KOALA (see Section 5.1.1) provide their own mechanism for the exchange of executable code: the HOC-SA supports the exchange of program pieces (code parameters) which are stored in the Code Service, while KOALA uses the Globus Resource Allocation Manager (GRAM [bC+98]), which supports the installation of a program on a remote host, by submitting the program (together with a description of it, in the Globus Resource Specification Language RSL) to a Web service. A detailed comparison of the HOC-SA with WS GRAM was given in Section 3.3.

Both code transfer technologies can be combined in a single application that installs the HOCs it needs on demand using WS GRAM and transfers code parameters to them using the Code Service. On-demand installations of HOCs are never repeated, but once a HOC is installed, the HOC is reused in any application that requires it and only new code parameters are transfered again, since the code parameters of a HOC vary from one application to another.

As explained in Section 5.3, the CoreGRID component model (GCM [cCN05]) simplifies the development of applications that make use of multiple HOCs composed together. Such applications are scheduled by applying the submission technique recursively to each single HOC in the composition, processing the global HOC first, and then down to inner ones (more details follow in Section 5.1.6).

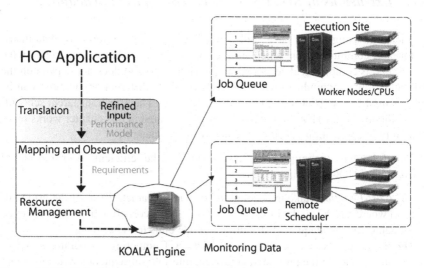

Figure 5.2 The 3-tier, User-Transparent Scheduling Infrastructure

5.1.3 Integrating KOALA & HOC-SA

For integrating KOALA with the HOC-SA, the 3-tier architecture shown in Fig. 5.2 was built. The shown tiers have the following functionality:

1. *Translation tier* for mapping the user's selection of a HOC to an associated communication pattern and the most appropriate cost-function. This tier is also responsible for identifying whenever the reusable scheduling technique can be used and repeated submissions of the same code can be avoided. It is represented by the server-site part of the HOC-SA where application code units are composed into sets of jobs (application code + RSL, see Section 5.1.1) and associated with an appropriate cost-function (see Section 5.1.5).

2. *Mapping and observation tier* for tracking allocated resource status and application progress and for taking action whenever a failure occurs and additional resources are needed. This tier is represented by the remote schedulers in the

---

Algorithm BWCF (J [n], S [N])
  /* n - the number of job stages; N - the number of sites */
  /* J [n] - the application stages; S [N] - the list of sites */
  1. order job stages according to their dependencies

      /* Build the set of mappings for the 1st stage */
  2. S_1 = set of potential execution sites for the first stage
  3. **if** ($S_1 \neq \emptyset$) **then**
  4.   **for each** (E ε S_1) **do**
  5.   estimate the file transfer cost TF_E
  6. Q_1.add (E , (-1, TF_E))
  7. **else**
  8.   **return** mapping failure (no available execution sites)

      /* Build the set of mappings for the jth stage */
  9. **for each** (job stage j from 2 to n) **do**
  10. S_j = set of potential execution sites
  11.   **if** ($S_j \neq \emptyset$) **then**
  12.   **for each** (E ε S_j) **do**
  13.   estimate file transfer $TF_{(E)}$ cost
  14.   **for each** (F ε elements (Q_{j-1})) **do**

          /* Estimate communication cost */
  15. $TC_{(F,E)} = Cost(E, AL (Q_{j-1}(F)))$

  16.   **if** $TF_{(F,E)} + TC_{Input,E} + Q_{j-1}[F] < C$ **then**
  17.   x = E and F_x = F
  18. $C = TF_{(F,E)} + TC_{I,E} + Q_{j-1}[F]$)
  19. Q_j.add (x, (F_x, C))
  20.   **else**
  21.   **return** mapping failure (no available execution sites)

      /* Extract the final HOC application mapping */
  22. **for** i = n **downto** 1 **do**
  23. N_i = E such that MIN (Q_j [F]) and F = Q_i[E]

Figure 5.3 The BWCF Algorithm for HOC Scheduling

KOALA infrastructure: e.g., the Sun Grid Engine [aC^{+}04] plus the Globus Monitoring and Discovery Service (MDS [bH^{+}05]).

3. *Resource management tier* for acquiring and aggregating adequate resources to fulfill user objectives. This tier implements the aggregated submission of code units and is represented by the KOALA Engine and the MDRunner, introduced in Section 5.1.1.

5.1.4 A HOC-Aware Scheduling Algorithm

The close-to-file policy (CF) [bME05], as the KOALA engine usually applies it to tasks submitted by any of the "classic" KOALA runners (KRunner, DRunner, GRunner), reduces the costs associated with transferring the input/output files needed for running an application. But CF does not take into account the communication requirements implied by a distributed parallel application.

The MDRunner, instead, uses an algorithm (shown in Fig. 5.3) that combines the close-to-file policy with communication cost awareness. This algorithm is called BWCF for BandWidth-aware and Close-to-File scheduling and was developed by Cătălin L. Dumitrescu [bD+06] during his CoreGRID fellowship on scheduling HOC applications. The main advantage of the BWCF algorithm, as compared to KOALA's greedy CF algorithm, is that it provides for applications with inter-process communication an optimal mapping (as formally proven [bE+6]).

In the worst case (each stage exchanges messages with all the others, which implies n-1 links for each stage), the complexity of BWCF is $\mathcal{O}(n^2 * N)$ due to line 15. The complexity of the BWCF algorithm depends though on the type of HOC being used and the number of available execution sites. For the Pipeline-HOC, it is $\mathcal{O}(n * N)$ for a number of n HOC instances scheduled over a number of N available sites.

5.1.5 HOC Scheduling Cost-Functions

To map HOC applications optimally to the available grid resources, the BWCF algorithm, as employed by the enhanced KOALA scheduler relies on a cost-function (see Fig. 5.3, line 15).

It is crucial to the BWCF algorithm that the cost function picked to compute the total communication costs (the value TC in line 16) reliably determines the minimum costs for each mapping. Therefore, the MDRunner (see Section 5.1.1) implements six different cost-functions, suitable for different applications [bE+6]. The four general ones are shown in Fig. 5.4.

1. **Link-Count-Aware:** In equation (1) in Fig. 5.4, N represents the number of neighbor units scheduled on the same site (one network link) and M the number of neighbor units scheduled on remote sites, which are weighed as three network links (3 was proven a reasonable factor in experiments). When applications are scheduled using this function, slower links are weighed with higher costs;
2. **Bandwidth-Aware:** In equation (2) in Fig. 5.4, LC_i is the bandwidth of the intra-site links used by the code unit i, and RC_j is the bandwidth of the inter-site links used by the code unit j, with N and M introduced before. This cost-function optimizes over the communication time required for each code unit to exchange information with all its peers. Also for this funtion is reliable and taking all peers

into account is not too much, as our experiments proof. Formula (3) sums up the required communication times for local and remote communication, taking network bandwidths into account;

$$
\begin{aligned}
&(1)\ F_{link\_count}(j) = N + 3 \times M \\
&(2)\ F_{bw\_aware}(j) = \sum_{i=1}^{N} LC_i / TBW + \sum_{k=1}^{M} RC_k / TBW \\
&(3)\ F_{lt\_aware}(j) = \sum_{i=1}^{N} LC_i / TLT + \sum_{k=1}^{M} RC_k / TLT \\
&(4)\ F_{net\_util}(j) = \sum_{i=1}^{N} LC_i^j / ABW_j + \sum_{k=1}^{M} RC_k^j / ABW_j
\end{aligned}
$$

Figure 5.4 General Cost-Functions for Estimating Communication Costs

3. **Latency-Aware:** In equation (3) in Fig. 5.4, TLT stands for the total network latency ($[sec]$). This cost-function optimizes over the latency encountered by each unit when exchanging information with all its peers. As in equation (2), the costs associated with each unit-to-node mapping are weighed by a different factor;

4. **Network-Load-Aware:** In equation (4) in Fig. 5.4, the bandwidth factor, which is a constant in the previous cost-functions, is replaced by the dynamically monitored bandwidth (ABW_j). In LC_i^j, the index j denotes the code unit number, i.e., the j-th unit has N local peers and M remote peers, while i stands for the index of the unit's neighbor.

The following two additional cost-functions were specifically designed for scheduling HOC applications:

$$
\begin{aligned}
&(5)\ F_{app\_aware}(j) = \sum_{l=1}^{j} \sum_{i=1}^{N} LC_i^l / ABW_j + \sum_{l=1}^{j} \sum_{k=1}^{M} RC_k^l / ABW_j \\
&(6)\ F_{variance}(j) = F_{app\_aware}(j, AL) - PredVar(link\_bw)
\end{aligned}
$$

Figure 5.5 HOC-Aware Cost-Functions

5. **Application-Aware:** The cost-function in equation (5) in Fig. 5.5 weighs local and remote link capacities, as well as the simultaneous communications for all units of an application (given by the employed HOC type and captured by the external summation). Thus, it is assumably the most appropriate one for applications with similar requirements for all units, like, e.g., most farm-based ones, while it is less appropriate for pipelined computations that benefit most from the Latency Aware cost-function (see experiments);

6. **Predicted-Network-Variance-Aware:** This cost-function in equation (6) in Fig. 5.5 incorporates measured network variances and is probably most appropriate for applications with specific requirements for each code unit. It optimizes

over both the time required to perform the communication between distributed units and the estimated communication penalties due to the network load produced by other applications. Since any change in the bandwidth usage of a HOC application is supposed to repeat in the future, the predictive part is computed as the average of the observed changes, as proposed in [bY+03].

By experiments [bE+6], it was verified that the most universal cost-function overall is the *bandwidth-aware* function, i.e., with the *bandwidth-aware* cost-function any application runs more effectively than with Close-To-File scheduling. By HOC type, the most appropriate functions are *application-topology-aware* for the Farm-HOC, and *latency-aware* for the Pipeline-HOC. For the Wavefront-HOC, the *bandwidth-aware* and the *predicted network variance* cost-functions lead to the most efficient scheduling. A summary of these experiments follows in Section 5.2.5.

5.1.6 Scheduling Large-Scale Applications

In order to schedule applications that use multiple HOCs, the communication costs for each single HOC must be summed up to obtain a scheduling criterion, which is valid for the combination of all the employed HOCs. This technique is used by KOALA for scheduling large-scale applications which are built by gluing multiple components together (e.g., via connecting component membranes, as discussed in Section 5.3 for the GCM [cCN05]) and require the computation of a complex schedule (i.e., one that involves more than one cost-function).

Reusing Complex Schedules

Computing a schedule for a combination of k HOCs has k times the complexity of the BWCF algorithm from Section 5.1.4 ($\mathscr{O}(n^2 * N)$), i.e., it is a quite costly operation. Since many different applications can be built using the same combination of components on the same host configuration, different applications using the same HOCs will be scheduled using the same mapping, i.e., the same distribution of code to grid hosts (provided that these hosts are still available). To avoid the need to compute the same schedules again and again, the MDRunner from Section 5.1.1 includes the support for reuse of schedules for already processed HOC compositions.

5.1.6.1 The Impact of uSLAs on HOC Execution

The management of access policies using uSLAs and its influence on grid application scheduling was extensively analyzed by Cătălin L. Dumitrescu [bE+6]. In large-scale application which are composed of multiple HOCs, the sharing rules under which resources are made available to the HOCs can be expressed using uSLAs. These uSLAs govern the sharing of specific resources among multiple consumers:

once a consumer is permitted to access a resource via an access policy, the resource uSLA steps in to govern how much of the resource can be consumed.

Some resource access limitations are not controlled by the scheduler, but via operating system programs (`xinetd`, `tcpd`, etc). In its standard implementation, the KOALA scheduler deals with access limitations via trial-and-error: when a KOALA Runner reports repeated failures of a job at a site (e.g., due to insufficient resources), the site is temporarily removed from the pool of available resources. The enhanced KOALA version for HOC scheduling (introduced in Section 5.1.2) implements a *fixed-limit* uSLA mechanism: the `MDRunner` ensures that the resource managers (see Section 5.1.3) are never overloaded, e.g., by limiting node usage to a single application at a time. A *space-shared* grid environment like the DAS-2 testbed [cDu06] guarantees this per default.

A more complex user rights and reservation priority management is beyond the scope of this book, but discussed, e.g., in [bWL07].

5.1.7 Experiments with HOCs and KOALA

The testbed used for evaluating the enhanced KOALA version for HOC scheduling was the DAS-2 platform [cDu06], a wide-area network of 200 Dual Pentium-III computer nodes. It comprises sites of workstations, which are interconnected by SurfNet, the Dutch University Internet backbone for wide-area communication, whereas Myrinet, a popular multi-Gigabit LAN, is used for local communication.

Several experiments were conducted to study the effects of automatizing the choice of execution nodes in HOC applications [bE+6]. The experiments demonstrated that the most efficient schedules for executing HOCs on multiple distributed grid sites (each comprising a multitude of nodes) can be determined in a user-transparent manner using the algorithm from Section 5.1.4 of this chapter.

We have performed three sets of experiments for three different scenarios in scheduling HOC-based applications, namely *communication-size analysis, cost function analysis,* and *schedule reuse analysis.* The results are captured in Tables 5.1, 5.2, and 5.3. The values in parentheses are the number of units and messages. The values in parentheses in the next two tables are the number of units, the number of exchanged messages and the size of each message.

Communication-Size Analysis

For the communication-size analysis, we perform 10 runs of each single HOC type and we present the average value and the standard deviation. Each HOC application was composed of 15–22 units and ran on three DAS-2 sites: Delft, Utrecht, and Leiden.

For the Farm-HOC, we observed similar performance when using *link-count* and the CF policy (see Table 5.1), which we explain by the identical mapping of units to resources. However, when more resources were available, KOALA's default

Table 5.1 Speedup for Link-Count (%)

Comm.	Synthetic HOC Application Type		
Req.	*Farm (15x20)*	*Pipe (15x20)*	*Wave (22x20)*
50Kb	0.71±3.2	10.95±6.6	15.81±15.2
100Kb	0.47±2.2	12.15±7.9	14.11±6.4
500Kb	0.87±4.5	14.68±2.5	14.22±4.6
1Mb	0.19±2.8	13.59±4.9	14.26±2.6
5Mb	0.89±1.4	12.56±5.5	15.38±2.8
10Mb	0.23±5.9	13.61±5.6	14.86±5.6

scheduling policy performs worse as shown in Table 5.2. For the Pipeline-HOC and the Wavefront-HOC, the performance increase when using the *link-count* is similar, regardless of the amount of input data (the gain is expressed in percentage for 20 data items of 50Kb–10Mb).

Cost Function Analysis

For the cost function analysis, each HOC application exchanged 20 messages with 2–10 Mb of data (a 10 times higher communication requirement than in the previous scenario), while running on all five sites of DAS-2.

Table 5.2 captures our results for running communication-intensive HOC-based applications: they consist of 20–50 units of work and have high communication requirements compared to the capacity of the inter-site network links. We observe that the *link-count* cost function (our previous choice) yields a lower performance than the *bandwidth-aware* cost function, while our implementation of the *application-aware* cost function introduces the lowest performance, even lower than the performance of the default CF scheduling policy (Table 5.2). The value for the standard deviation is quite high (Table 5.2), since we compared BWCF also with KOALA's default scheduling policy, which leads to much higher time needs in component-based applications, regardless of the employed cost function.

Table 5.2 Speedup for the Six Costs (%)

	Synthetic HOC Application Type		
Cost Function	*Farm* (20×20×2M)	*Pipeline* (50×20×10M)	*Wavefront* (46×20×10M)
Link Count	7.18±9.6	21.46±19.4	34.00±26.9
Bandwidth	34.06±6.8	20.90±19.5	39.37±29.2
Latency	31.06±2.8	24.60±19.2	28.30±29.2
Network	33.80±6.8	22.60±19.1	17.40±33.6
Application	36.10±5.9	-5.00±0.1	-1.13±4.6
Predicted	34.21±1.8	21.04±19.5	38.22±6.2

In summary, the most appropriate cost function seems to be the *bandwidth-aware* function. By HOC type, the most appropriate functions are *application-aware* for the Farm-HOC, and *latency-aware* for the Pipeline-HOC, which supports our assumption. For the Wavefront-HOC, the *bandwidth-aware* and *predicted* cost functions lead to the most efficient scheduling.

Schedule Reuse Analysis

In the schedule reuse analysis, we test the performance of our scheduling when schedules are reused for HOC-based applications that exhibit the same workflow, i.e., they employ the same HOCs in identical order. Also, a *fixed-limit* uSLA at the user level was enforced. The uSLA allowed at most four applications to run in parallel on the DAS-2 resources. Table 5.3 captures our results for HOCs with 15–22 units and message sizes of 1Mb–10 Mb.

Table 5.3 Throughput Gains with Reuse (%)

Cost Function	Synthetic HOC Workload Type		
	Farm 20×30×1M	Pipeline 20×30×10M	Wavefront 20×30×5M
CF	93.7	112.9	101.8
Link	100.7	116.3	161.0
Bandwidth	118.6	126.4	170.4
Latency	117.4	133.6	161.6
Network	106.3	134.8	170.2
Application	101.0	117.4	127.3
Predicted	118.2	123.6	194.0

We note a high throughput improvement due to the schedule reuse: the reduction of the scheduling overhead allows to increase the total throughput by more than 100% in our test scenario.

This observation can be easily traced in Fig. 5.6, where the work unit termination time is plotted on the vertical axis. Once a HOC instance terminates and is requested again, the reserved resources are reused instead of computing a new schedule.

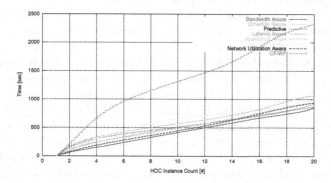

Figure 5.6 Farm-HOC Execution Shapes

The experimental results show an increase in throughput with more than 100%, a lowering of the response time by 50%, and a failure reduction by 45%, as compared to a Close-To-File scheduling which is usually chosen, either by an automatized job to grid node mapper (e.g., the classic KAOLA version) or a human user.

5.1.8 Conclusions from the Scheduling Experiments

In this section, the user-transparent scheduling approach for Higher-Order Components was introduced and analyzed from two perspectives, namely (1) the easiness of integration within the already existing KOALA scheduling infrastructure, and (2) the performance gains. The presented approach combines several scheduling techniques, like the cost-based reducing of the communication times, the reuse technique for avoiding the need of repeated scheduling for the HOCs (or combinations of HOCs as workflows) and the aggregated submission technique.

The approach was evaluated by integrating the HOC-SA (from Chapter 3) with the KOALA grid scheduler [bME05], as described in Section 5.1.3 (including the listed extensions for user-transparent scheduling), and by running different test applications on the DAS-2 testbed combining over 200 nodes at multiple sites in the Netherlands.

5.2 Parallelization of Code Parameters in HOCs

This section describes a combination of two research areas of parallel and distributed computing: (a) component-based grid programmingusing HOCs and (b) automatic parallelization of program loops. In this combination, HOCs enable the reuse of component code in application development. When a programmer is provided, e.g., with the Farm-HOC, only the independent worker tasks can be supplied as code parameters. But, once an application exhibits data or control dependences, the trivial farm is no longer sufficient. Here, the power of loop parallelization tools, like LooPo [cUP97], comes into play: the combination of HOCs with LooPo facilitates the automatic transformation of a sequential loop nest with complex dependences (supplied by the user as a HOC parameter) into an ordered task graph which can be processed on the grid in parallel.

The parameters of a HOC often carry a loop nest to be executed by some worker hosts (i.e., any free processing nodes) in the grid. A typical example is the Farm-HOC from Chapter 2. The original Farm-HOC is not able to deal with inter-task dependences: they would make it necessary either to design a new HOC which takes the dependences into account or to remain with a sequential, less efficient solution. Instead of requiring the developer to build one new HOC per possible dependence pattern, in this section, a more flexible component, called LooPo-HOC, which embeds the LooPo loop parallelizer into the Farm-HOC, is suggested. An

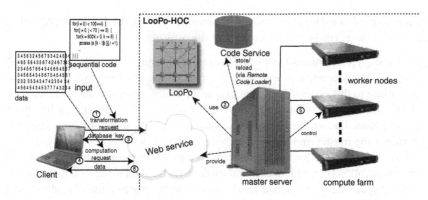

Figure 5.7 General Setup of the LooPo-HOC in the HOC-SA

equation system solver based on the successive overrelaxation method (SOR) is used as a motivating application example for the LooPo-HOC and for performance experiments.

The LooPo-HOC is composed of an internal compute farm implementation for running the actual application tasks (Section 5.2.1, Fig. 5.7, right), LooPo itself for transforming code (Section 5.2.2, Fig. 5.7, middle), the Web service for clients to connect and controller software for task queue management and workload monitoring (Section 5.2.3). These parts are available to the client via a single Web service (Fig. 5.7, left).

Figure 5.7 shows how the LooPo-HOC works on a high level:

① The client issues a transformation request via the Web service that the master provides as access point to the LooPo-HOC.
② The code parameter that the client sends as input to the master, together with the request, is a nest of sequential loops.
③ Transparently to the client, the uploaded code parameter is transformed by LooPo into a graph structure, wherein independent tasks are grouped together.
④ The Web service returns the key for referencing to the transformed code which is stored in the Code Service and can be executed in parallel using an adapted task farm.
⑤ The client sends this key together with the application input data as parameters of the computation request to the Web service.
⑥ This request is processed in parallel by the compute farm formed by the worker nodes.
⑦ Finally, the result data is sent back to the client.

5.2.1 The Internal Compute Farm of the LooPo-HOC

Since LooPo resolves inter-task dependences, the parallel processing on the grid is handled using a compute farm and the LooPo-HOC can be viewed as an extended version of the Farm-HOC[bA$^+$07]. The compute farm in the LooPo-HOC differs from that in the Farm-HOC in Chapter 2 in two ways:

1. the LooPo-HOC embeds LooPo and uses it for ordering tasks in the form of a task graph taking dependences among tasks into account. The farm executes the tasks according to this order, freeing the user from dealing with task dependences.
2. the communication is more efficient in the LooPo-HOC, especially in fine-grain applications, i.e., applications where many small data units are communicated between the master and the worker nodes, since a Web service is only used as the remote interface for supplying input via an Internet connection. All internal communication (between master and workers) is handled using a communication library, specifically developed for the LooPo-HOC (by Eduardo Argollo during his CoreGRID fellowship on automatic loop parallelization for HOCs [bA$^+$07]). Practically, this library is a light-weight version of MPI, supporting all the basic and most of the collective operations, using only Java and TCP sockets.

As the Farm-HOC, the LooPo-HOC offers a universal farm implementation, i.e., this farm is capable of executing applications without dependences as well (and has shown almost linear speedup in various experiments). It is included in the open-source distribution of the HOC-SA (introduced in Chapter 3) in the package `farmService.mwDependence`.

The worker nodes in Fig. 5.7 are fully decoupled from each other, i.e., they need no communication between each other, and are supposed to run in a distributed environment. The following subsection describes in more detail the transformation process, the scheduling and the workload monitoring, which make up the core of the LooPo-HOC and which are supposed to run locally, on one server, ideally on a multiprocessor machine. This server has three different responsibilities:

1. providing a public access point to the LooPo-HOC in form of a Web service,
2. running LooPo for processing transformation requests,
3. running the compute farm (i.e., controlling the worker nodes) for processing computation requests.

As illustrated in Fig. 5.7, only the worker nodes and the Code Service host (see Chapter 3) are physically separate machines. But principally, the above three responsibilities can also be shared among multiple servers, which especially makes sense when many clients issue concurrent requests and no multiprocessor server is available.

5.2.2 Transforming Loop Nests into Task Graphs

For the automatic parallelization of loop nests using LooPo, there are three passes involved, to transform the input loop to a parallel target program in which the granularity of parallelism is adapted to the number of available grid nodes and, independently, to the cost ratio of computation vs. communication [bG$^+$04].

Phase 1: Program and dependence analysis

The loop parallelization methods used here are applicable to perfectly or imperfectly nested loops which compute on array variables and whose bounds and data dependences are affine expressions, i.e., linear in the indices of the surrounding loops and in symbolic and numeric constants. The dependence analysis is based on the *polytope model* [bLe93]. To test whether any two array accesses, for possibly different values of the loop counters, reference the same cell in this model, LooPo tests inequations corresponding to the surrounding loops. Using LooPo, the LooPo-HOC can determine which instance of which statement most recently accessed the same array cell [bCG99].

Phase 2: Parallelization

Space-time mapping is a technique used to extract parallelism from a loop nest. The parallelism is expressed by (piece-wise) affine functions mapping every instance of every statement to coordinates in space, i.e., on different virtual processors, and in time, i.e., to different logical execution times. These functions are called *placement* and *schedule* [bCG99]. The schedule has to respect the dependences; computations that are scheduled at the same logical time can be executed in parallel. The placement is done by a heuristics for locating instances of statements depending on each other on the same virtual processor.

Phase 3: Granularity adaptation

In most cases, the granularity of the parallelism generated in Phase 2 is much too fine to be efficient. Thus, tiling techniques are applied to the space-time mapped program. Tiling means aggregating a set of virtual processors. All virtual processors that are covered by one tile are executed on the same physical processor. The tile size must be a compromise between much, but fine-grained and, thus, communication-intensive parallelism on the one hand, and little parallelism but also a small communication volume on the other. A cost model can be used to adjust the size with the ratio of communication cost and computation cost.

Each tile produced by LooPo represents a *task* for the LooPo-HOC and contains the corresponding set of computation operations for the time steps and virtual processors that were aggregated. Information about data dependences between tasks is stored in the form of a *task graph* that is used by the master for scheduling them, i.e., for choosing an order of execution between dependent tasks.

Mapping Tiles to Processors

The master is responsible for arranging the execution order, whereas the target processor for the execution can be determined using an advanced scheduling system, e.g., the KOALA scheduler from Section 5.1.1, to exploit task locality. In grid environments which do not provide a scheduling system with tunable policies (such as KOALA [bD+06]), users of the LooPo-HOC can also directly adapt the master, such that the complete scheduling is handled there. This way, programmers can, e.g., arrange chains of tasks that should be executed on the same worker. For data dependences between tasks that make the exchange of computed data elements necessary, the master provides a method to *join* a new (dependent) task with a finished task. This way, the dependent task is decoupled from its predecessor, gets the updated data and is scheduled for execution.

For many input loops, the load in the generated parallel program is unevenly distributed among the processors. Frequently, the parallel program first has a phase of increasing parallelism, and afterwards, sometimes immediately following, a phase of decreasing parallelism. Thus, if all existing parallelism is exploited, i.e., if as many processors are used as there can be used maximally, there is theoretically a maximal efficiency of 50%. The only way to solve this problem is to reduce the number of processors used. Tiling alone cannot improve this situation: affine functions, which are the mathematical basis of tiling, cannot map different tiles to the physical processors in a round-robin fashion, or in any other way so that the number of tiles to be mapped to a physical processor changes with the available parallelism. However, this kind of load balancing is one of the strengths of task farming as it is performed by the LooPo-HOC using the workload monitor described below.

5.2.3 Integrating Loop Parallelization with the Grid

Beside the workers (executing the single tasks, as described in Section 5.2.1) and the master (running LooPo, as described in Section 5.2.2), the LooPo-HOC comprises a Web service. This Web service provides a new interface to LooPo for remotely accessing it in the grid. Besides the public remote interface, this Web service includes a resource configuration for maintaining the distributed application state (status data and intermediate results), as is typical in the Web Service Resource Framework (WSRF) [bH+05].

The LooPo-HOC preserves task graphs across multiple sessions using WS resources, which are referenced by endpoints (the *resource keys*) and reused in multiple applications. The LooPo-HOC also benefits from the support for asynchronous operations in Globus (WS-N [cOA06]). While LooPo transforms loop nests, the client can disconnect or even shut down. The LooPo-HOC can restore the task graph from a former session, when the client sends to it the corresponding resource key. The LooPo-HOC uses two types of WS resources. For every code transformation request, one new resource instance (i.e., transient storage) for holding the resulting task graph is created dynamically. The other resource is static (i.e., instantiated only once and shared globally among all processes), called *monitor* and explained below.

The task graph resources are instantiated following the factory pattern [aG+95], returning the unique resource key to the client. As shown in Fig. 5.8, the client

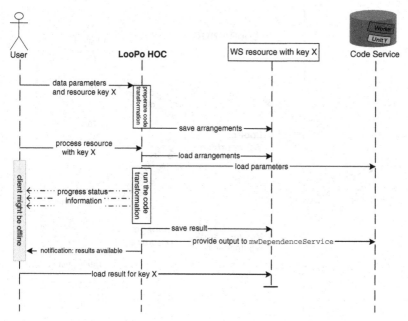

Figure 5.8 Sequence Diagram for Using the LooPo-HOC

sends the resource key on every communication with the LooPo-HOC, which uses the key afterwards to retrieve the corresponding resource data (the task graph and intermediate results, stored persistently). Thus, the master server in Fig. 5.7 is not a single point of failure, but rather a service provider that permits clients to switch between mirror hosts during a session.

Workload Monitoring in the LooPo-HOC

The transformation of loop nests into tasks graphs is a computation-intensive operation, which is quite unusual for Web services: typically, a Web service operation retrieves or joins some data remotely and then it terminates. Due to the asynchronous operations of the LooPo-HOC, the clients produce processing load right *after* their requests are served, since this is the moment when the code transformations begin (concrete time costs follow in Section 5.2.5).

From the user's viewpoint, the asynchrony is advantageous, since local application activities are never blocked by the code transformations. However, when multiple users are connected to the same LooPo-HOC server, the workload must be restricted to a certain (server-specific) maximum of concurrent requests. For this purpose, the LooPo-HOC has a workload monitor (see Fig. 5.9) which provides status information to the clients.

The workload monitor consists of two parts, a fixed-size thread pool and a status map (bottom right in Fig. 5.9). For every transformation, the LooPo-HOC first

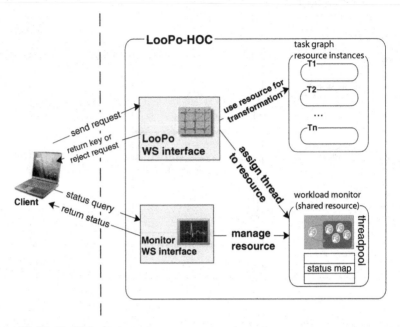

Figure 5.9 Workload Monitor

checks if an idle thread is available. If the thread pool is fully loaded, then the LooPo-HOC creates a new transformation thread and adds it to the pool. The maximum threshold for the thread pool is set by the server administrator and is usually equal to the number of CPUs of the hosting server. Once the number of executing threads has reached this maximum, new incoming requests are queued.

The status map (shown down right in Fig. 5.9) is a structured data store, used to keep track of the successive transformations. The client can read the map by issuing an XPath query [cW399] to the monitor at any time. This feature is useful when the client reconnects during a loop transformation. The map also allows one application to execute the tasks resulting from transforming the sequential loops submitted by another application: via the map, users can track the status of transformations (and run the resulting tasks), even if they connect to the LooPo-HOC for the first time (and, consequently, receive no automatic notification about status updates). This scenario arises, e.g., if the Web service for connecting to the LooPo-HOC is deployed to multiple servers, allowing clients to switch between hosts when a connection fails or some host's request queue is full.

5.2.4 Case Study: The SOR Equation System Solver

As an example application for the LooPo-HOC, this section shows a solver for linear equation systems, $A\phi = b$, using the successive overrelaxation method

(SOR). The SOR method works by extrapolating the Gauss-Seidel method [aSa03], as shown in the following iteration formula:

$$\phi_i^{(k+1)} = (1-\omega)\phi_i^{(k)} + \frac{\omega}{a_{ii}}\left(b_i - \Sigma_{j=1}^{i-1}a_{ij}\phi_j^{(k+1)} - \Sigma_{j=i+1}^{n}a_{ij}\phi_j^{(k)}\right)$$

Here, vector $\phi^{(k)}$ denotes the kth iterate, the a_{ij} are elements of the input matrix A, and ω is called the *relaxation factor* which is reduced in each iteration until it declines below some tolerance. Roughly speaking, the SOR algorithm computes weighted averages of iterations, leading to a triangular system of linear equations, which is much simpler to solve than the original arbitrary system. There is one control dependence in the SOR solver, i.e., in each pair of successive iterations, the follow-up statement depends on its predecessor.

To run this application in parallel on the grid, the user supplies the LooPo-HOC with the application name (here, SOR) and a sequential description of the computations expressed in Java notation, as shown in Fig. 5.10.

```
1: @LooPo("begin loop", "constants: m,n; arrays: a{n+1}")
2: for (int k = 1;  k <= m;  k++)  {
3:     for (int i = 2;  i <= n-1;  i++)  {
4:         // average computation
5:         a[i] = (a[i - 1] + a[i + 1]) / 2.0; ... } }
6: @LooPo("end loop")
```

Figure 5.10 The Sequential Code Parameter

Any loop nest (with metadata, as the annotations in lines 1 and 6) can be used as input for the LooPo-HOC. The automatic parallelization steps from Section 5.2.2 are performed to obtain a task graph. First, a model of the input program is derived (Fig. 5.11). For the space-time transformation, the following schedule θ and placement π were determined:

$\theta(k,i) = 2*k+i$ and $\pi(k,i) = k$.

For obtaining the task dependence graph, tiling is applied on the transformed target program. Figure 5.12 shows the representation of the transformed program after tiling is applied using the same color tone for tiles that can be executed independently. The final model is derived by joining each tile as a node into a graph with every inter-tile dependence as a directed edge in that graph.

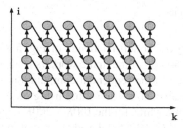

Figure 5.11 The Input Program (M=7, N=5)

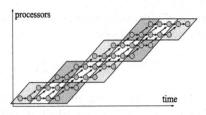

Figure 5.12 Transformed Program After Tiling

From the task graph model representation, the LooPo-HOC generates three Java classes, holding the transformed code, which are stored in the Code Service: SORMaster, SORData and SORTask. The SORMaster holds the dependence graph (it implements the interface SchTaskDependence from the mwDependenceService package) and provides the join-method (required for distributing data to task groups; see Section 5.2.2). SORData objects are used for buffering application data and the SORTask class describes a single task (as an implementation of the execute-method from the interface mwDependenceService.UserTask).

Since these files comply with the interface definitions in the HOC-SA [cD⁺06], the user can directly load the three output files for parallel processing as code parameters of the farm described in Section 5.2.1.

5.2.5 Experiments

Figure 5.13 shows the computation times for matrices of different sizes using from 1 to 10 workers running on 1.7 GHz PCs in the University of Münster computer pool with a data throughput of approximately 3.4MBit/s.

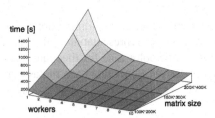

Figure 5.13 Computation Times

The strong time decay on the left-hand side of Fig. 5.13 (from 1 to 3 workers) shows that especially the adding of the first 2 workers leads to a strong performance improvement, as compared to the sequential computation time. The decline of the plane along the z axis (matrix size) shows that using more than 5 workers is only beneficial for large matrices, while, for the $100K \times 200K$ matrix there are not enough tasks (using a 5×5 tiling [bG⁺04]) to take advantage of more than 4 workers.

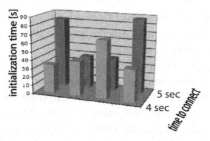

Figure 5.14 Initialization Times

The eight bars in Fig. 5.14 represent the initialization times for 10 workers (i.e., the time that passes by after the client sends a request, until the remote computations in the farm start). The time required to establish an `ssh` connection between the master and all workers varied between 4 and 5 seconds. As can be observed by comparing the bars in the front row and the back row, there is no correspondence between the time required to connect and the full initialization time (including the remote code loading), which exhibits strong variations between 30 and 90 seconds (the standard deviation σ from the mean value of 50 seconds was 22). This is due to the connection between the farm and the database: as explained in Section 5.2.1, the farm workers load the code for processing the single tasks from the Code Service using OGSA-DAI [cUD07], which is known to deliver unreliable performance under certain conditions, especially when it is deployed on a single server together with other Web services [bH+06]. In relation to the much longer computation times of the SOR application (from several minutes to several hours for large matrices), the initialization time can be disregarded. It should also be noted that the initialization is only performed once per worker and application. After the first set of input data (a matrix in the SOR example) has been processed, the same parallel code is used to process any number of successive inputs without repeating its generation (using LooPo) and its transfer from the Code Service to the workers.

The transformation of the single loop nest used in the SOR example in Section 5.2.4 takes approximately 1 min on a contemporary dual-core PC, utilizing 50 % of its overall CPU capacity. From this quick increase of computational load, it can be concluded that, if only one server is used to run the code transformations in multiple different applications of the LooPo-HOC, then this machine should be a powerful multiprocessor server.

5.3 Combining HOCs with Related technologies: ProActive, SOFA and the GCM

The HOC approach to the development of grid software consists in combining high-level constructs, which simplify application programming, with Web services, which enable interoperability among distributed components. The high-level constructs are developed as HOCs either from scratch, e.g., the Farm-HOC in Chapter 2, or they are developed using a general-purpose library, e.g., *eSkel* as used in the Lifting-HOC in the previous section.

Many different grid programming tools were developed independently from the HOC approach to grid programming, e.g., the popular ProActive library [cPA07] and the implementation of the CoreGRID Grid Component Model (GCM [cCN05]) based on ProActive. As mentioned in Chapter 2, Higher-Order Components (HOCs) are themselves one possible implementation version of the CoreGRID GCM. In this section, the combination of HOCs with the ProActive-based GCM implementation is considered, allowing the HOC developer to benefit from the special features of the ProActive library.

Features of ProActive/GCM

One simplification which developers can achieve by using the ProActive-based GCM version in combination with HOCs, is that ProActive/GCM provides support for composing hierarchical components out of other components (e.g., by nesting HOCs into each other). For this reason, the use of ProActive/GCM for implementing new HOCs as assemblies of smaller-grained components is studied in this section.

The ProActive library offers (amongst many other features [cPA07]) also a generator for automatically creating Web services for accessing components which adhere to the GCM standard (or its predecessor *Fractal* [cFr99]), as long as all the parameters used by these components are primitives. An important question studied in this section is, if a HOC can be made remotely accessible via an automatically created Web service in the same manner. The following problem arises: HOCs have both primitive and – in the form of code parameters – complex parameters, as well. How can the complex parameters of a HOC automatically be translated into their valid representation (using the corresponding identifiers, as explained in Chapter 2) in the Web service interface?

This section shows how this problem is solved by combining the HOC remote code-loading mechanism with the ProActive library. The employed technologies (ProActive/GCM) are also described in more detail.

Features of HOCs as Skeletons

A promising idea for simplifying development and enhancing application quality is skeleton-based development [aCo89], as used for the Lifting-HOC in the preceding section. This approach is based on the observation that parallel applications share a common set of recurring patterns such as divide-and-conquer, farm, and pipeline. The idea is then to capture such patterns as generic software constructs that can be customized by developers to produce particular applications. HOCs can be viewed as skeletons implemented as components which are exposed via Web services. The Web services allow any Internet-connected client to access HOCs and request from them the execution of standard parallelism patterns on the grid.

HOCs meet the requirements which a grid component model, according to the CoreGRID GCM specification (GCM [cCN05]), must satisfy with respect to abstraction and interoperability: (1) the skeletal programming model offered by HOCs imposes a clear separation of concerns in terms of high-level services expecting from their users the provision of application-level code only, and (2) any HOC offers a publicly available interface in form of a Web service making it accessible from remote systems without introducing any requirements on them, e.g., regarding the use of a particular middleware technology or programming language.

Building new grid applications using HOCs is simple as long as they require only HOCs that are readily available (or can be built by adapting the available HOCs). However, once an application requires a specific parallelism pattern that is not covered by the available HOCs, a new HOC has to be built. For building HOCs, the Service API can be used (Section 2.3) but this API still requires from the programmer to configure Web service interfaces and WS resources [cOA04] manually, which, especially for complex components, is tedious and error prone. As shown in the following, the HOC developer can be freed from manually configuring Web services for accessing HOCs which are built as combinations of ProActive/GCM components.

5.3.1 Combining HOCs with ProActive, the GCM and SOFA

To demonstrate that HOCs and the ProActive-based GCM version are complementary, an alternative version of the Farm-HOC from Chapter 2 was built (shown in Fig. 5.15) using the ProActive-based GCM implementation.

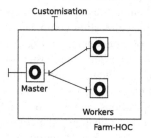

Figure 5.15 Farm-HOC Shown Using GCM Symbols

The interface Customisation of the ProActive/GCM-based Farm-HOC in Fig. 5.15 corresponds to the original Farm-HOC interface, but in the ProActive/GCM version, the Java code is different (Fig. 5.16). The ProActive/GCM Component-type is used for master and worker, i.e., both parameters are components of their own allowing the user, e.g., to define another Farm-HOC and nest it into a worker. To make the ProActive/GCM-based farm component accessible as a HOC in the grid, the Customisation interface must be exposed via a Web service.

However, this requires that one can pass a code-carrying, behavioral argument (e.g., the `master`) to this Web service. Moreover the service must be associated with state data, such that the customizations triggered by the `setMaster/setWorker`-operations have a persistent effect on the HOC.

```
public interface Customisation {
    public void setMaster(Component master);
    public void setWorker(Component worker);
}
```

Figure 5.16 Farm-HOC Interface in the ProActive/GCM Version

Service state data for HOCs is declared in the appropriate configuration files (Chapter 2) and for the ProActive/GCM-based Farm-HOC, the configuration of the original Farm-HOC from Chapter 2 can be reused. Although the parameters in the `Customization`-interface in Fig. 5.16 are declared using the ProActive `Component`-type (instead of `int`-IDs), the ProActive/GCM-based Farm-HOC offers the same public operations taking the same number of arguments as in the original Farm-HOC. As shown below, also the `Component`-parameters can be specified using `int`-IDs (the only difference is that ProActive has a special referencing mechanism for code [cPA07]).

The Hierarchical Component Composition Features

A distinctive feature of the ProActive-based GCM implementation is its support for hierarchical component composition: each `Component`-instance offers an interface called *membrane* which simplifies combining it with other components: there are many library methods for connecting component membranes, e.g., for building assemblies and nestings of components. In Fig. 5.15, the membranes were not shown, since there is one for each item, the master, the workers and the composite. The ProActive-based GCM implementation includes support for an XML-based architecture description language (ADL) which allows users to describe, besides interfaces, hierarchical compositions, and also to modify such compositions (e.g., exchange subcomponents) without changing the component implementation. This feature was adapted from the Component Definition Language (CDL) of the SOFA component model [bP+98]. This mechanism described in Section 5.3.2 allows HOC applications to benefit from the features of SOFA and ProActive as well.

Even without the ProActive-based GCM implementation, the combination of multiple HOCs in one application is generally possible, since clients can issue as many requests to HOCs as an application requires (and even a HOC's code parameters may contain calls to other HOCs). However, the HOC Service API from Section 2.3 provides no support for building new HOCs by combining existing ones. When developers build applications that use multiple HOCs, they must explicitly deal with each of them in the code they write, when only the HOC Service API is used. The component nesting and combination features of the ProActive-based GCM implementation can fairly simplify the development of such applications: recurrently used component assemblies, such as, e.g., a farm of pipelines can be once connected via the component membranes and reused in many different applications (provided that a version of each of the employed components based on ProActive/GCM exists). As a rule of thumb, it makes sense to combine HOCs with the ProActive-based GCM version, whenever a large number of components is used in a single application or components are deeply nested into each other.

The Web service creation mechanism in ProActive/GCM can not automatically build service configurations for a grid middleware like Globus. However, HOCs are preconfigured to work with this type of middleware. In the following, it is shown how the code transfer mechanism of HOCs can be integrated into ProActive/GCM for automatically creating Web services that allow the interconnection of distributed grid components which exchange data and code over the network.

The Automatically Generated Remote Interface

This section describes the existing ProActive mechanism for automatically exposing GCM components as Web services, and then it explains how this mechanism was extended with the code mobility mechanism of HOCs.

ProActive uses the Axis [cAp03b] library to generate WSDL descriptions and the Apache SOAP [cAp03a] engine to deploy Web services automatically. Service invocations are routed through a custom ProActive *provider*. When a GCM component should be exposed as a Web service, the ProActive user simply calls the static library method `exposeComponentAsWebService`, which generates the required service configuration and makes a new Web service available. The URL of this new service is specified as a parameter.

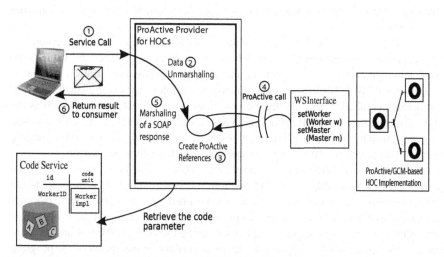

Figure 5.17 ProActive/GCM Web Services and HOC Code Loading

When a call (Fig. 5.17, step ①) reaches such a service, the Apache SOAP engine unmarshals the request message and forwards it to the corresponding ProActive provider for serving the request (step ②). For enabling the use of code parameters, a special ProActive provider for HOCs was developed [bB$^+$06]. This provider accepts code parameters (e.g., for the master or for the workers of the ProActive/GCM-based Farm-HOC in Fig. 5.15) in the form of primitive identifiers. The provider retrieves the code referenced by the identifiers from the Code Service and creates a ProActive reference for each of them (Fig. 5.17, step ③) using the `Component`-constructor from the ProActive library [cPA07]. Then, the Provider issues a standard ProActive call to the HOC using the ProActive references as parameters (step ④) and returns the result to the SOAP engine. The engine then marshals a new SOAP message (step ⑤) and sends it back to the service consumer, i.e., to the client (step ⑥).

5.3.2 Creation of Web Services Using ProActive

This section presented a solution to supporting code-carrying parameters in automatically created Web services for accessing grid components that were developed as compositions of HOCs and ProActive/GCM-based grid components.

Further work is to devise a general automatic solution for supporting arbitrary components and code parameters, even a complex Java type, when publishing components via Web services. This section presents a first step in this direction: it suggests a solution that only applies to some specific parameter types, i.e., those representing *behaviors* which are expressed as HOC code parameters. Behaviors are represented by code which is not contained inside the component implementation (or, at least, not fixed there: HOCs can provide default behaviors). But a behavior is not an arbitrary code: it must adhere to a fixed interface (e.g., the `Master` which must provide a `split`-method). The general case would call for a solution where the generation of Web service interfaces would be automated for any code-carrying parameter.

For addressing the general case, related work should be taken into account: (1) the *valuetype* construct in CORBA, which supports passing objects by value (both state and behavior) to remote applications [cCO04], (2) possible – not yet standard – extensions of WSDL for passing arguments of complex types using specific SOAP Attachments [aEr04], and (3) standard facilities for XML data binding, such as the Java Architecture for XML Binding JAXB [cSM04]. Whatever the solution used for passing parameters of arbitrary types is, it calls for a generic and automatic mechanism based on reflection techniques and dynamic code generation.

Note that legacy programs are another example of complex parameters which can be passed to a Web service. In Section 3.4, it was shown that the server-side HOC implementation can be, e.g., a legacy MPI program. But also a code parameter can be an MPI program. To run this program on top of the appropriate supporting environment (`mpirun` in the example case) using the correct libraries and data parameters (supplied to the program in the correct order), a description of the program must be shipped together with its code. The standard format for such descriptions in the context of Globus is the Resource Specification Language used by the Globus Resource Allocation Manager Web Service (WS-GRAM [bC$^+$98]). Chapter 3 explains the HOC-SA in more detail and compares it to WS-GRAM, showing the relation between both technologies.

5.4 Discussion: HOCs and Different Tools for Distributed Computing

Using the SOR program from Section 5.2.4 as an example, it was shown that the LooPo-HOC provides a promising scalability and that the time needed for the initial code transformations does not critically impact the overall application performance.

Another approach to automating the generation of parallel code was developed within the recent research on OpenMP programs and *reparallelizing* them for the grid [bK$^+$07]. This work also covers Java programs and the use of *Distributed Shared Memory* (DSM) for data exchange among tasks, but still requires from programmers dependence-free input and the explicit declaration of parallel loops via OpenMP directives. The LooPo-HOC, in contrast, offers a fully transparent programming interface that can run any sequential loop nest (including loops with complex dependences) in parallel. The required data sharing could have been implemented using Sun's standardized DSM implementation in *JavaSpaces* [cSM07e]. However, the LooPo-HOC requires only the joining of single tasks and no support for distributed transactions, and, thus, relies on a more light-weight messaging library [bA$^+$07], which can be supposed to provide better performance due to less overhead.

Since a Web service is used for accessing the LooPo-HOC remotely, it can easily be connected, e.g., to any GCM application [cCN05] which requires an efficient implementation for running parallel loops on multiple grid hosts. The suggested combination of components with loop parallelization is not only useful for the GCM, but also for other popular component models, such as CCA [cCC05] and CCM [cCO07]. Beside the HOC-SA code transfer mechanism (Chapter 3), no other special features of this or any other particular component technology are required.

This chapter has demonstrated that the integration of HOCs with other tools for distributed programming can further simplify the development of grid application and increase their performance. It was shown how the KOALA scheduler [bME05] can be integrated with HOCs for a user-transparent, automated choice of execution machines [bD$^+$06] (Section 5.1). While scheduling with KOALA increases the performance of HOCs by automatically adapting the execution environment, i.e., switching between resources, LooPo (Section 5.2) helps to reduce execution times by automatically adapting the application code.

Chapter 6
Conclusions and Related Work

This final chapter summarizes the main contributions of the book in Section 6.1. Section 6.2 is about related work. Many grid programming tools (libraries, schedulers, etc.) which are complementary to the HOC approach were already discussed in the previous chapters. This chapter presents a survey of the most popular component models for programming grid and Web applications, showing where HOCs and the HOC-SA range within the variety of contemporary approaches to grid programming. Moreover, this chapter suggests useful combinations of HOCs with Problem Solving Environments (PSEs) and exemplifies this class of software by presenting the most notable representatives of PSEs for grid programming. In the remainder of this chapter, the relation between the grid and other contemporary Internet-technologies such as the Semantic Web and the Web 2.0 is discussed and possible applications of HOCs which use such technologies are suggested. Finally, possible enhancements of the HOC model and the HOC-SA are suggested for future work in Section 6.3.

6.1 New Contributions

The present book makes the following new contributions:

1. **The Introduction of Higher-Order Components**
   HOCs are a novel type of software components (see Chapter 2) that simplify grid application programming via abstraction as follows: the parameters of a Higher-Order Component include the application-specific code of a user program, and only this code must be provided by the user. While grid programmers traditionally were responsible for all the technical details of their applications, HOCs hide many of these details inside the component implementation. Besides the general advantages of software components (separation of concerns and code reuse [aSz98]), HOCs are specifically useful in the grid context, as they shield the user from dealing with the complex communication issues and the required

J. Dünnweber, S. Gorlatch, *Higher-Order Components for Grid Programming*,
DOI 10.1007/978-3-642-00841-2_6, © Springer-Verlag Berlin Heidelberg 2009

middleware support in grid environments (as discussed in Chapter 1). Using some simple introductory examples, it was shown how to use HOCs (e.g., for computing fractal images). Chapter 4 shows how to apply HOCs for solving real-world problems on the grid (genome data processing and a crash test simulation) and presents performance experiments on grid testbeds.

2. **The Implementation of the HOC-Service Architecture**
   The HOC-SA (introduced in Chapter 3) is a Service-Oriented Architecture (SOA [aEr04]) that enables the use of HOCs on top of modern grid middleware, namely, the popular Globus Toolkit. The HOC-SA provides means for code transfer (the Code Service, the Remote Code Loader and a comfortable Web portal) which enable the exchange of code pieces written in Java or script languages which could traditionally not be exchanged over the network so easily. Chapter 3 presents a gateway for including native code in other languages (e.g., C/C++ or Fortran into Java). The application examples (e.g., in Chapter 4) show that HOCs work independently of a specific execution platform (grid testbeds, such as DAS-2 [cDu06] and PlanetLab [bP$^+$02] can be used) provided that the platform supports the use of Web services for interconnecting components and clients. Web services are currently the communication technology that enables interoperability among the broadest possible range of hardware architectures, and this book presents several enhancements to standard Web services (provided by the HOC-SA which, e.g., improves the performance of Web service-based applications and enables the use of code parameters).

3. **The Combination of Various Grid Programming Tools**
   This book shows that there are many complementary developments in the current research on grid computing. It is demonstrated how HOCs and the HOC-SA can be integrated with many of the other recently developed grid technologies, such that applications benefit from multiple tools. Examples for tools that can be advantageously combined with HOCs are: the KOALA scheduler [bME05] and the LooPo loop parallelizer [cUP97] (see Chapter 5), skeleton libraries like Lithium/μSkel [bA$^+$05] and eSkel [bCo04], ProActive [bB$^+$03] and the GCM/Fractal framework for hierarchical component composition [cCN05] (see Chapter 3).

6.2 Related Work

Many different high-level approaches to grid application programming and their relation to HOCs and the HOC-SA have been outlined in the corresponding sections of this book.

In this section, the distinct features of the multiple grid programming approaches are summarized, and suggestions are made, how users can combine HOCs and the related technologies for achieving synergistic effects.

Complex interconnections of multiple components are sometimes described as workflows and different, so-called problem solving environments (PSEs) have been designed for grid application development using workflows and graphical workflow languages. Notable examples are Cactus [bT+00], NAREGI PSE [bK+05], the Science Experimental Grid Laboratory SEGL [bC+05] and the Parallel Grid Run-time Application Development Environment P-GRADE [bK+06]. In a PSE supporting a graphical workflow language, programmers are not directly concerned with code, but only with software components which are represented by graphical symbols. For building applications, these symbols are interactively put together in a GUI which provides programmers with a very high level of abstraction over the underlying grid system. Obviously, this abstraction advantage comes at the price of reduced control over the grid system, since the developers must rely on the provided components and have little influence on the processes executed by them. Combining PSEs with the HOC-SA, where pieces of user-defined code can be used for customizing and adapting components, seems to be a promising idea toward combining PSEs with component programming. The HOC-SA portal, presented in Chapter 3 can be viewed as a step into this direction. All parts of this portal are independent Web pages which can be easily connected to a Web-based PSE by inserting the corresponding hyperlinks.

6.2.1 The Real-Time Framework (RTF)

Frequent interactions between distributed grid hosts serving vast user communities are studied in the *edutain@grid project* [bFa07] using challenging, real-time online games and simulation-based eLearning as example applications. The Real-Time Framework (RTF [bG+08]) – a novel middleware developed within the scope of this project – includes a variety of features which are of high value for grid computing, e.g., data replication for failure management, etc.

Aims of combining RTF with HOCs and even cloud computing technologies (see Section 3.5) can be twofold:

1. RTF frequently requires to allocate new resources when applications need to cope with a very high (and dynamically changing) number of users. Cloud computing networks like IBM's *blue cloud* [cFe08] are an interesting target platform to support dynamic resource (re)allocation. Especially because the cloud computing resources provide QoS guarantees and are, thus, more reliable than grid resources. Moreover, the cloud-internal protocol is not necessarily SOAP, such that RTF's own, much more efficient communication mechanisms can be employed.

2. Since RTF includes a sophisticated management layer with load balancing, etc., it can offer its own cloud services: access to RTF servers is typically managed in a service-oriented manner [bFa07], thus applications can make use of both, RTF and the HOC-SA, and the middleware also offers resource virtualization facilities.

6.2.2 A Survey of Related Component Models

This section contains a survey of the HOC-SA and other contemporary component programming technologies for grid and Web applications. To start this survey, the common definition of the term *software component* is briefly revisited.

Szyperski's popular definition of components requires that a component is a software entity "...with contractually-specified interfaces and explicit context dependences only" [aSz98]. In an implementation where components are self-contained software entities, the "context" is simply the target operating system. In this case, the system-level support for most applications is only rudimentary, because operating systems are usually designed universally, except some special cases, e.g., operating systems for embedded devices. When a more specific runtime environment, e.g., IBM WebSphere [cIB03a], is used for a certain component type, the "context dependences" increase to that effect that applications require exactly this environment. However, such a specific runtime environment can also provide more supporting functionalities for applications. Examples of software components that take advantage of a specialized runtime environment can be found in e-business systems. Some of these components are, e.g., automatically synchronized with databases, enterprise resource planning (ERP) or customer relationship management (CRM) software by the runtime environment (a special container). Programmers access these components via remote interfaces that abstract over the technical implications on other software and offer the components as high-level application entities, e.g., accounts or shopping carts.

One novelty of the HOC-SA is that any component (HOC) written on top of it implicitly uses a grid-enabled Web service container (Globus, in the current implementation) for handling the network communication. This brings the following advantages: interoperability with any technology that can access Web services; a loosely-coupled infrastructure where components can easily be exchanged, which facilitates fault tolerance; and finally, the implicit usage of standardized and secure Internet protocols for communication (HTTP and HTTPS). HOCs are closely related to the skeleton model [aCo89]. However, the use of grid middleware (the Globus container) in the HOC-SA introduces the new aspects discussed in this book (resource management, state data and the transfer of mobile code) which have so far never been studied in skeleton research. In the following, the relations between skeletons, other components and HOCs are discussed in detail.

6.2.3 The Skeleton Model

One motivation for *Higher-Order Components* (HOCs) is that in parallel and distributed computing, it is often useful to parameterize a component not only with data but also with application-specific code. This idea originates from algorithmic skeletons [aCo89], where the parallelization strategy is captured in code blocks (called *skeletons*) wherein the application-specific tasks are omitted and are expressed separately by the users as parameters. A parallelization strategy may describe a distributed evaluation schema for, e.g., divide-and-conquer, a pipeline or a farm. An application-specific task might be a step in a graph coloring, data tree traversal or sorting algorithm. By parameterizing a skeleton with code units, its strategy can be customized for a particular application. Contrary to HOCs, skeletons are self-contained programs which can be directly executed, once all parameters are defined; typically, skeletons are not deployed into containers and consequently no middleware configuration is necessary. However, many skeletons have an analogous HOC representation (see Section 3.2.3). The divide-and-conquer skeleton, e.g., and the corresponding Divide-and-Conquer-HOC both take parameters, such as the typical subroutines of the classical Quicksort algorithm [bHo62] for applying comparisons and subdividing according to a given pivot element. Both can then be used for parallel sorting, either as a stand-alone program (the skeleton) or as a component on top of a grid middleware (the HOC). For both constructs, the programmer provides only the application-specific code, while the distribution of data and computation for parallel processing is done inside the readily provided code.

Conceptually, a HOC can be viewed as a skeleton, made available in the grid as a remotely accessible component (using Web services and resources, as explained in Section 3.2). The required shipping of code over the network is implemented using the HOC-SA code mobility mechanism (as explained in Section 3.1.1).

6.2.4 CCA: The Common Component Architecture

CCA has been defined by a group of researchers from academia and industry committed to defining a standard component architecture for high-performance computing. The basic definition of a component in CCA [cCC05] states that a component "is a software object, meant to interact with other components, encapsulating certain functionality or a set of functionalities. A component has a clearly defined interface and conforms to a prescribed behavior common to all components within an architecture. Multiple components may be composed to build other components." Interfaces and behaviors in CCA are described by so-called *uses* and *provides ports*, which make components interoperable. In CCA, the systems that a component can interoperate with must support the format used to represent the component's ports. The HOC-SA, in contrast, implies the use of standard Web service interfaces. The CCA-model gives a very broad definition of software components, which is quite similar to Szyperski's definition [aSz98]. The HOC-SA can be viewed as an implementation of CCA for the grid. But (as it is the case for the CCM, discussed below) CCA does not provide its own component programming paradigm (such as the skeletal programming paradigm for HOCs in the HOC-SA). Current activities of the CCA forum include the definition of a CCA-specific format of component interfaces (Babel/SIDL) and framework implementations (Ccaffeine). The definition of components given in the OGSA-glossary [cGl96] is closely related to that in the CCA-glossary, in the sense that it also requires a *uses* and a *provides*-port for every component.

6.2.5 CCM: The CORBA Component Model

CCM is a component model defined by the Object Management Group (OMG) that produces and maintains computer industry specifications (e.g., CORBA, UML, XMI). The CCM specifications include a Component Implementation Definition Language (CIDL), the semantics of the CORBA Components Model (CCM), a Component Implementation Framework (CIF), which defines the programming model for constructing component implementations, and a container programming model. CCM is therefore technically similar to the HOC-SA, but the HOC-SA does not define any language of its own and CCM does not address any of the specific challenges of the grid.

6.2.6 Java Servlets and JSPs

Servlets and Java Server Pages (JSPs [cSM99], which are used for generating Servlets dynamically from tagged page-templates) process HTTP-requests by rendering HTML or XML data. Servlets are compositional units with explicit interface and dependency definitions and, therefore, Servlets are components as well. They also rely on a container that binds operation requests to implementation code, like Web services. The difference between Servlets and Web services is that Servlets are used for handling requests that were submitted by human users via Web browsers, while Web services are used for inter-machine communication.

Also the Globus Container relies on the Java Servlet technology for maintaining services. Users can browse and manage the services deployed to a host via a Web browser using an adapted version of the Apache Axis Servlet [cAp03b]. Thus, a full runtime environment for Servlets and JSPs is available on every server in a Globus-based grid. The HOC-SA exploits this feature in the user-friendly portal (built upon Servlets, JSPs and Apache Struts [cAp03c], see Chapter 3) which allows users to submit code parameters to a HOC using their standard Web browser.

6.2.7 Enterprise Java Beans and .NET Components

In contrast to the HOC-SA model, targeted mostly towards compute-intensive applications, Sun Microsystems' EJBs were defined for distributed e-commerce applications. Therefore, EJBs focus on transaction-, security- and database-management [aMo06], similarly to Microsoft's competing business component framework .NET [aLo01].

The Globus Toolkit contains some scripts that allow to expose the so-called home interface of an EJB automatically as a service in the grid [cTT04]. A home interface is used to create EJB-instances. So, if a HOC-SA application should benefit from the features of an EJB (e.g., high availability via pooled instances), a HOC can request an EJB-instance from the associated home interface and delegate its invocation via SOAP to the Globus container. A similar approach to exposing .NET components in the grid using Globus has been adopted in the MyCoG.NET project, a specialized Commodity Grid (CoG) API for programming grid applications with Microsoft .NET [bP+07].

There are many similarities between WSRF [cOA04] and the business components in .NET and the EJB model. Most notable are probably *life-cycles* (i.e., automatic callbacks which the container triggers upon user-defined events, such as a new resource creation [cOA04]) and *resource homes*. Resource homes are designed according to the factory pattern [aG+95] and located via JNDI, exactly like an EJB home interface.

Contrary to the ten different HOC types listed in Section 3.2.3, the EJB model defines only three basic component types (from which all other types are derived by inheritance): (1) *session beans* for transient application entities, such as shopping carts which clients can instantiate and destroy on-demand; (2) *message-driven beans* for application activities such as inventory updates which users define for running them asynchronously with other application activities (standard Java multi-threading can not be used for describing asynchronous activities in an EJB system, since thread control is taken over by the container); and (3) *entities* for persistent objects such as customer accounts. An extensive description of example use cases for EJBs in a realistic e-commerce system can be found in [bV+02]. The .NET framework emerged from Microsoft's COM+ technology and offers similar components for composing e-business applications (using a different terminology). An extensive description of .NET and example applications can be found in [aLo01]. Microsoft's .NET also supports multiple programming languages and even introduces new ones (C# and J#).

Compared to such mature component architectures, it is questionable whether Web services themselves should be equated with components, although this classification is not wrong, according to [aSz98] (and sometimes suggested, e.g., in [bG+03]). However, Web services are very simple components. In contrast to EJBs or .NET components, Web services offer merely an access mechanism, while it is left open to the programmer what is accessed via this mechanism. Web services are software entities that benefit from container functionalities (e.g., the automatic encoding of input and the decoding of output data), but they still require the programmers to implement all operations which they make accessible over the network themselves. The purpose of the HOC-SA is to integrate Web services with components (HOCs) that implement the most commonly used processing patterns to simplify grid application programming and enable a high degree of network interoperability, at the same time.

Besides the grid, which is sometimes cited as the Internet technology of the future [aFK98], two other Internet-based developments currently attract a rising interest: the Web 2.0 [cOR05] and the Semantic Web [bB+01].

One common idea behind these two research directions and the grid technology this book deals with is simplicity through transparency.

6.2.8 The Web 2.0

Web 2.0 applications are more responsive as compared to traditional Web applications since the interactions are more direct, e.g., instead of typing absolute numbers into forms, users drag around intuitive graphical entities (boxes, pencils, scissors, magnifying glasses, etc.). Such improvements are made possible by decomposing Web applications into many small pieces which can be distributed, such that only that pieces are placed on remote servers which cannot be placed locally on the client (e.g., because they directly operate on remote data). Most parts run directly on the client, leading to improved responsiveness for the users. Such decompositions are obvious to the Web developers, who are concerned with so-called portlets, instead of one monolithic portal. Web 2.0 users are shielded from the details of where the multiple parts of a complex online transaction take place. A notable example application is Google Maps [cGM05] giving users the impression of a fully local application when navigating through a map from a server-side database via moving inside or zooming in and out of it with the mouse.

6.2.9 The Semantic Web

In the Semantic Web, data is annotated and structured in ontologies that associate data with a meaning. An often-cited example for the benefits of the Semantic Web is that of a Web service developer who searches material about the Web service protocol SOAP: submitting a query for "SOAP" to a traditional Web search engine will probably result in hits that contain material on hand soap, dish detergents and soap operas. In a Semantic Web application, in contrast, the developer can conduct the search for "SOAP" only in the context of network communication, allowing the search engine to automatically filter out irrelevant material and also include information on related topics such as WSDL, UDDI, etc. As an example application, the Semantic Open Data Tool [cCh06] allows users to perform a Web search returning hits that contain the users' exact search keywords but also related material which belongs into the same model (declared, e.g., using the Resource Description Framework RDF [cW304]), since crosslinks are automatically established.

Due to the use of Web services for communication, the HOCs presented in this book can be easily combined with modern Web technology. The Web 2.0 technology *Asynchronous JavaScript and XML* (AJAX [cMD08]), e.g., has been successfully used for making the HOC portal highly responsive (see Section 3.2.4). The use of semantic information (see Section 3.1.2) for classifying code parameters can be viewed as an integration of HOCs with the Semantic Web. This is not only useful for searching code parameters but, e.g., the LooPo-HOC (Section 5.2) may also benefit from such a classification, since it may help to avoid the repeated transformation of code parameters which have different names but the same semantics. Similar efforts are currently being made in the e-Science and the Semantic grid community [cSo07].

The support of script languages for parameterizing HOCs (see Chapter 3), corresponds to the recent paradigm shift in Web application development, from Java-based frameworks like Apache Struts [cAp03c], Java Server Faces [cSM05a] and Spring [cSP08] toward the use of script languages for customizing existing code, such as, e.g., in the popular Ruby On Rails (RoR) [cRR07] framework. While the former frameworks heavily involved programmers in the use of XML for configuring their applications, RoR is an alternative providing so-called "scaffolding" which helps to construct Web applications much easier. The notions which are often cited in the context of RoR are "convention over configuration" and "agile development" and also apply to the use of HOCs and code parameters for grid programming.

6.3 Future Work

The convenience of grid programming and the quality of grid solutions will decide the future of grids as a new Internet technology. As shown in this book, the current state of the art in application programming for the grid, with middleware systems positioned between the system and the programmer, is not satisfactory. Grid application programmers are distracted from the main task they are experts in – developing efficient and accurate algorithms and services for demanding problems – and have to invest most of their effort into low-level, technical details of a particular middleware.

HOCs were proposed as a novel component model for bridging the gap between grid middleware and grid application programmers.

Besides the readily available features in the HOC-SA implementation (separation of concerns, compositionality and code reuse) HOCs could be enhanced further in the following aspects:

- **Formalism/Abstraction** can be facilitated for HOCs due to their mathematically well-defined nature. Components are formally defined as higher-order functions, with precise semantics, which allow to develop formal rules for semantics-preserving transformations of single components and their compositions in the programming process [bGo96].
- **Performance prediction** is an important and challenging problem for grid programming. HOCs as well-defined parallel programming constructs with pre-packaged implementations have a strong potential for developing analytical cost models [bAG02]. A pipeline-structured component can, e.g., be represented using a process algebra for computing a numerical estimation of the expected data throughput [bB$^+$04]. Process algebra and other formal models, such as Petri Nets, can lead to a cost calculus [bA$^+$06] which facilitates performance prediction of grid applications built of HOCs.

The success of future-generation grid systems will strongly depend on whether these systems can be programmed easily and efficiently by a wide community of

Bibliography

References

(a) Textbooks

[aB⁺91] Carl-Ivar Branden, John Tooze, and Carl Branden. *Introduction to Protein Structure.* Garland Science, 1991.

[aC⁺04] Jason Carolan, Paul Strong, Ed Turner, and Scott Radeztsky. *Building N1 Grid Solutions: Preparing, Architecting, and Implementing Service-Centric Data Centers.* Sun BluePrints Program. Prentice Hall, 2004.

[aCo89] Murray I. Cole. *Algorithmic Skeletons: A Structured Approach to the Management of Parallel Computation.* Pitman, 1989.

[aEr04] Thomas Erl. *Service-Oriented Architecture: A Field Guide to Integrating XML and Web Services.* Prentice Hall, 2004.

[aFK98] Ian Foster and Carl Kesselmann, editors. *The Grid: Blueprint for a New Computing Infrastructure.* Morgan Kaufmann, 1998.

[aG⁺95] Erich Gamma, Richard Helm, Ralph Johnson, and John Vlissides. *Design Patterns: Elements of Reusable Object-Oriented Software.* Addison Wesley, 1995.

[aG⁺02] Ananth Grama, Anshul Gupta, George Karypis, and Vipin Kumar. *Introduction to Parallel Computing.* Addison-Wesley Longman Publishing Co., Inc., 2002.

[aGR93] Jim Gray and Andreas Reuter. *Transaction Processing: Concepts and Techniques.* Morgan Kaufmann, 1993.

[aHu98] Barbara Burke Hubbard. *The World According to Wavelets: The Story of a Mathematical Technique in the Making.* AK Peters Ltd., 1998. Second ed.

[aJH01] Arne Jensen and Anders la Cour-Harbo. *Ripples in Mathematics: the Discrete Wavelet Transform.* Springer, 2001.

[aKa03] Doug Kaye. *Loosely Coupled: The Missing Pieces of Web Services.* Rds Associates Inc, 2003.

[aLa03] Ramnivas Laddad. *AspectJ in Action.* Manning, 2003.

[aLo01] Juval Lowy. *COM and .NET Component Services.* O'Reilly & Associates, Inc., 2001.

[aMB90] Tony Mason and Doug Brown. *lex & yacc.* O'Reilly & Associates, Inc., 1990.

[aMo06] Richard Monson-Haefel. *Enterprise Java Beans 3.0.* O'Reilly and Associates, 2006.

[aMu08] James Murty *Programming Amazon Web Services,* O'Reilly, 2008

[aPH04] Jack Park and Sam Hunting. *XML Topic Maps: Creating and Using Topic Maps for the Web.* Addison-Wesley, 2004.

[aPL91] Przemyslaw Prusinkiewicz and Aristid Lindenmayer. *The Algorithmic Beauty of Plants.* Springer, 1991.

[aPR96]   Heinz-Otto Peitgen and Peter H. Richter. *The Beauty of Fractals, Images of Complex Dynamical Systems*. Springer-Verlag, 1996.

[aRG03]   Fethi A. Rabhi and Sergei Gorlatch, editors. *Patterns and Skeletons for Parallel and Distributed Computing*. Springer, 2003.

[aSa03]   Yousef Saad. *Iterative Methods for Sparse Linear Systems*. SIAM USA, 2003.

[aSB99]   Luis Silva and Rajkumar Buyya. *High Performance Cluster Computing: Programming and Applications*, edited by Rajkumar Buyya. Prentice Hall, 1999.

[aSz98]   Clemens Szyperski. *Component Software: Beyond Object-Oriented Programming*. Addison Wesley, 1998.

[aV$^+$02] Markus Völter, Alexander Schmid, and Eberhard Wolff. *Server Component Patterns. Component Infrastructures Illustrated with EJB*. John Wiley & Sons, 2002.

(b) Research Papers

[bA$^+$02] Martin Alt, Holger Bischof, and Sergei Gorlatch. Algorithm Design and Performance Prediction in a Java-based Grid System with Skeletons. In Burkhard Monien and Rainer Feldmann, editors, *Euro-Par 2002*, pages 899–906, Springer (LNCS 2400), Aug. 2002.

[bA$^+$04] Jan Dünnweber, Martin Alt, and Sergei Gorlatch. APIs for Grid Programming using Higher-Order Components. In Thilo Kielmann Simon J. Cox and Stephen Pickles, editors, *GGF12 - The Twelfth Global Grid Forum, Brussels, Belgium*, Sept. 2004.

[bA$^+$05] Marco Aldinucci, Marco Danelutto, Jan Dünnweber, and Sergei Gorlatch. Optimization Techniques for Skeletons on Grids. In L. Grandinetti, editor, *Grid Computing and New Frontiers of High Performance Processing*, Advances in Parallel Computing, pages 255–273. Elsevier, Nov. 2005.

[bA$^+$06] Martin Alt, Sergei Gorlatch, Andreas Hoheisel, and Hans-Werner Pohl. Using High-Level Petri Nets For Hierarchical Grid Workflows. In *2nd IEEE International Conference on e-Science and Grid Computing*, Amsterdam, Dec. 2006.

[bA$^+$07] Eduardo Argollo, Michael Claßen, Philipp Claßen, and Martin Griebl. Loop Parallelization for a Grid Master-Worker Framework, *CoreGRID* Tech. Report 80, pages 516–527, June 2007.

[bAG02]   Martin Alt and Sergei Gorlatch. Performance Prediction of Java Programs for Metacomputing. In *5th International Workshop: The Internet Challenge – Technology and Applications*, pages 111–119. Kluwer, Sept. 2002.

[bAG04]   Martin Alt and Sergei Gorlatch. Adapting Java RMI for Grid Computing. *Future Generation Computer Systems*, 21(5):699–707, May 2004.

[bAn02]   Anderson David, Jeff Cobb, Eric Korpela, Matt Lebofsky and Dan Werthimer. SETI@Home: An Experiment in Public-Resource Computing. *Communications of the ACM*, 45(11):56–61, Nov. 2002.

[bB$^+$01] Tim Berners-Lee, James Hendler, and Ora Lassila. The Semantic Web. *Scientific American*, 284(5):34–43, May 2001.

[bB$^+$02] Laurent Baduel, Françoise Baude, and Denis Caromel. Efficient, Flexible, and Typed Group Communications in Java. In *Java Grande Conference*, pages 28–36, Seattle, ACM, Nov. 2002.

[bB$^+$03] Françoise Baude, Denis Caromel, and Matthieu Morel. From Distributed Objects to Hierarchical Grid Components. In *International Symposium on Distributed Objects and Applications (DOA)*. Springer LNCS 2888, Catania, Sicily, Nov. 2003.

[bB$^+$04] Anne Benoit, Murray Cole, Stephen Gilmore, and Jane Hillston. Evaluating the Performance of Pipeline-Structured Parallel Programs with Skeletons and Process Algebra. In Frédéric Loulergue, editor, *Practical Aspects of High-level Parallel Programming*, pages 299–306. Springer, June 2004.

[bB⁺06]   Jan Dünnweber, Sergei Gorlatch, Françoise Baude, Virginie Legrand, and Nikos
           Parlavantzas. Towards Automatic Creation of Web Services for Grid Component
           Composition. In Sergei Gorlatch and Marco Danelutto, editors, *Integrated Research
           in GRID Computing*, pages 31–42. Springer, Dec. 2006.
[bBE01]   Anca I. D. Bucur and Dick H. J. Epema. The Influence of Communication on the
           Performance of Co-Allocation. In *7th International Workshop on Job Scheduling
           Strategies for Parallel Processing*, pages 66–86, London, UK, Springer, June 2001.
[bBu02]   Janusz M. Bujnicki. Sequence Permutations in the Molecular Evolution of DNA
           Methyltransferases. *BMC Evolutionary Biology*, 2:3, Mar. 2002.
[bC⁺98]   Karl Czajkowski, Ian Foster, Carl Kesselman, Stuart Martin, Warren Smith, Steven
           Tuecke. A Resource Management Architecture for Metacomputing Systems. In *IPP-
           S/PDP'98 Workshop on Job Scheduling Strategies for Parallel Processing*, pages 62–
           82, Mar. 1998.
[bC⁺05]   Natalia Currle-Linde, Uwe Küster, Michael Resch, and Beneditto Risio. Science Ex-
           perimental Grid Laboratory (SEGL) Dynamic Parameter Study in Distributed Sys-
           tems. In *PARCO*, vol. 33 of *John von Neumann Series*, pages 49–56. Institute for
           Applied Mathematics, Jülich, Germany, Sept. 2005.
[bC⁺06]   Jan Dünnweber, Sergei Gorlatch, Anne Benoit, and Murray Cole. Component-Based
           Grid Programming. A Case Study on Wavelets. In HPC EUROPA Report *Science and
           Supercomputing in Europe*, pages 215–218, CINECA, Dec. 2005.
[bCa07]   Reed A. Cartwright. Ngila: Global Pairwise Alignments with Logarithmic and Affine
           Gap Costs. *Bioinformatics*, 23(11):1427–1428, Mar. 2007.
[bCG99]   Jean-François Collard and Martin Griebl. A Precise Fixpoint Reaching Definition
           Analysis for Arrays. In Larry Carter and Jeanne Ferrante, editors, *Languages and
           Compilers for Parallel Computing, 12th International Workshop, LCPC'99*, pages
           286–302, Springer (LNCS 1863), Aug. 1999.
[bCo04]   Murray I. Cole. Bringing Skeletons out of the Closet: A Pragmatic Manifesto for
           Skeletal Parallel Programming. *Parallel Computing*, 30(3), Mar. 2004.
[bD⁺02]   John Anvik, Steve MacDonald, Duane Szafron, Jonathan Schaeffer, Steve Bromling,
           and Kai Tan. Generating Parallel Programs from the Wavefront Design Pattern. In
           High-Level Parallel Programming Models held in conjunction with the *International
           Parallel & Distributed Processing Symposium* (IPDPS-16). IEEE, Apr. 2002.
[bD⁺03]   Alexandre Denis, Christian Perez, and Thierry Priol. Padico: An Open Integration
           Framework for Communication Middleware and Runtimes. *Future Generation Com-
           puter Systems*, 19(4):575–585, Apr. 2003.
[bD⁺04]   Cătălin L. Dumitrescu, Ioan Raicu, Matei Ripeanu, and Ian Foster. DiPerF: An Au-
           tomated Distributed Performance Testing Framework. In *GRID '04: Proceedings of
           the Fifth IEEE/ACM International Workshop on Grid Computing (GRID'04)*, pages
           289–296, Pittsburgh, PA, USA, IEEE, Nov. 2004.
[bD⁺05]   Jan Dünnweber, Sergei Gorlatch, Anne Benoit, and Murray Cole. Integrating MPI-
           Skeletons with Web Services. In *PARCO*, vol. 33 of *John von Neumann Series*, pages
           787–794. Institute for Applied Mathematics, Jülich, Germany, Sept. 2005.
[bD⁺06]   Cătălin L. Dumitrescu, Dick H.J. Epema, Jan Dünnweber, and Sergei Gorlatch. User-
           Transparant Scheduling of Structured Parallel Applications in Grid Environments. In
           HPC-GECO/CompFrame Workshop at HPDC-15, Paris, France, IEEE, June 2006.
[bD⁺07]   Cătălin L. Dumitrescu, Jan Dünnweber, Philipp Lüdeking, Sergei Gorlatch, Ioan
           Raicu, and Ian Foster. Simplifying Grid Application Programming Using Web-
           Enabled Code Transfer Tools In *Towards Next Generation Grids*, pages 225–235.
           Springer, Aug. 2007.
[bDC04]   Guillaume Dewaele and Marie-Paule Cani. Interactive Global and Local Deforma-
           tions for Virtual Clay. *Graphical Models*, 66(6):352–369, June 2004.
[bDG04]   Jan Dünnweber and Sergei Gorlatch. HOC-SA: A Grid Service Architecture for
           Higher-Order Components. In *International Conference on Services Computing
           (SCC04), Shanghai, China*, pages 288–294, IEEE, Sept. 2004.

[bDS04]   Jeffrey Dean and Sanjay Ghemawat. MapReduce: Simplified Data Processing on
            Large Clusters, In *Proceedings of the 6th Conference on Symposium on Operating
            Systems Design & Implementation, San Francisco, CA*, Pages 10–10, Dec. 06–08,
            2004.

[bDG05]   Jan Dünnweber and Sergei Gorlatch. Component-based Grid Programming Using the
            HOC-Service Architecture. In Issam Hamido Fujita, editor, *New Trends in Software
            Methodologies, Tools and Techniques*, Frontiers in Artificial Intelligence and Appli-
            cations. IOS Press, Sept. 2005.

[bE⁺6]   Cătălin L. Dumitrescu, Dick H.J. Epema, Jan Dünnweber, and Sergei Gorlatch.
            Reusable Cost-Based Scheduling of Grid Workflows Operating on Higher-Order
            Components. In *2nd IEEE International Conference on e-Science and Grid Com-
            puting*, Amsterdam, Dec. 2006.

[bFa07]   Thomas Fahringer, Christophe Anthes, Alexis Arragon, Arton Lipaj, Jens-Müller-
            Iden, Chris Rawlings, Radu Prodan and Mike Surridge. The edutain@Grid Project,
            In *4th International Workshop on Grid Economics and Business Models*, 2007.

[bFl72]   Michael J. Flynn. Some Computer Organizations and Their Effectiveness. In *Com-
            puters*, C-21(9):28–33, IEEE, Sept. 1972.

[bFo06]   Ian Foster. Globus Toolkit Version 4: Software for Service-Oriented Systems. In *IFIP
            International Conference on Network and Parallel Computing*, pages 2–13. Springer,
            Oct. 2006.

[bG⁺76]   Jim Gray, Donald Chamberlin, Michael W. Blasgen, and Morton M. Astrahan. A
            Relational Approach to Database Management. *ACM Transactions on Database Sys-
            tems*, 1(2):97–137, June 1976.

[bG⁺01]   Patricia Gonzalez, Jose Cabaleiro, and Tom Pena. Parallel Computation of Wavelet
            Transforms Using the Lifting Scheme. *Journal of Supercomputing*, 18(2):141–152,
            Feb. 2001.

[bG⁺03]   Sanjay Gosain, Arvind Malhotra, Omar El Sawy, and Fadi Chehade. The Impact of
            Common e-Business Interfaces. *Communications of the ACM*, 46(12):186–195, Dec.
            2003.

[bG⁺04]   Martin Griebl, Peter Faber, and Christian Lengauer. Space-Time Mapping and Tiling
            – a Helpful Combination. *Concurrency and Computation: Practice and Experience*,
            16(3):221–246, Mar. 2004.

[bG⁺06]   Jan Dünnweber, Sergei Gorlatch, Sonia Campa, Marco Danelutto, and Marco Aldin-
            ucci. Adaptable Parallel Components for Grid Programming. In Sergei Gorlatch and
            Marco Danelutto, editors, *Integrated Research in GRID Computing*, pages 43–57.
            Springer, Dec. 2006.

[bG⁺08]   Sergei Gorlatch, Frank Glinka, Alexander Ploss, Jens Müller-Iden, Radu Prodan, Vlad
            Nae, and Thomas Fahringer. Enhancing Grids for Massively Multiplayer Online Com-
            puter Games. In *Euro-Par 2008*, University of Las Palmas, Gran Canaria, Spain, Aug.
            2008.

[bGD05]   Sergei Gorlatch and Jan Dünnweber. From Grid Middleware to Grid Applications:
            Bridging the Gap with HOCs. In *Future Generation Grids*. Springer, Nov. 2005.

[bGo82]   Osamu Gotoh. An Improved Algorithm for Matching Biological Sequences. *Journal
            of Molecular Biology*, 162:705–708, Dec. 1982.

[bGo96]   Sergei Gorlatch. From Transformations to Methodology in Parallel Program Devel-
            opment: A Case Study. *Microprocessing and Microprogramming*, 41:571–588, Apr.
            1996.

[bGo04]   Sergei Gorlatch. Send-Receive Considered Harmful: Myths and Realities of Message
            Passing. *ACM TOPLAS*, 26(1):47–56, Jan. 2004.

[bH⁺90]   Xiaoqiu Huang, Ross C. Hardison, and Webb Miller. A Space-Efficient Algorithm for
            Local Similarities. In *Computer Applications in the Biosciences*, volume 6(4), pages
            373–381. Oxford University Press, Oct. 1990.

[bH⁺98]   Jian Huang, Roni Yagel, Vassily Filippov, and Yair Kurzion. An Accurate Method
            for Voxelizing Polygon Meshes. In *IEEE Symposium on Volume Visualization*, pages
            119–126, Oct. 1998.

Bibliography

175

[bH+02]   Chafia Hejase de Trad, Qiang Fang and Irena Cosic. Protein Sequence Comparison Based on the Wavelet Transform Approach. In *Protein Engineering 15*. Oxford University Press, Mar. 2002.

[bH+05]   Marty Humphrey, Glenn Wasson, Jarek Gawor, Joe Bester, Sam Lang, Ian Foster, Stephen Pickles, Mark Mc Keown, Keith Jackson, Joshua Boverhof, Matt Rodriguez, and Sam Meder. State and Events for Web Services: A Comparison of Five WS-Resource Framework and WS-Notification Implementations. In *14th IEEE HPDC-14*, July 2005.

[bH+06]   William Hoarau, Sébastien Tixeuil, Nuno Rodrigues, Décio Sousa, and Luis Silva. Benchmarking the OGSA-DAI Middleware. In Thierry Priol Sergei Gorlatch, Marian Bubak, editor, *Proceedings of the CoreGRID Integration Workshop*, pages 357–368. CYFRONET Poland, Oct. 2006.

[bHo62]   Charles Antony Richard Hoare. Quicksort. *Computer Journal*, 5(1):10–15, British Computer Society, Apr. 1962.

[bJe99]   Albert Jeltsch. Circular Permutations in the Molecular Evolution of DNA Methyltransferases. *Journal of Molecular Evolution 49*, pages 161–164, July 1999.

[bK+02]   Jens Kleinjung, Nigel Douglas, and Jaap Heringa. Parallelized Multiple Alignment. In *Bioinformatics 18*. Oxford University Press, Sept. 2002.

[bK+03]   Nicholas T. Karonis, Brian Toonen, and Ian Foster. MPICH-G2: a Grid-Enabled Implementation of the Message Passing Interface. *Journal of Parallel and Distributed Computing*, 63(5):551–563, May 2003.

[bK+05]   Hiroyuki Kanazawa, Motohiro Yamada, Yutaka Miyahara, Yoshikazu Hayase, Shigeo Kawata, and Hitohide Usami. Problem Solving Environment based on Grid Services: NAREGI-PSE. In *1st IEEE International Conference on e-Science and Grid Computing*, Washingtion, pages 456–463, Dec. 2005.

[bK+06]   Peter Kacsuk, Tamas Kiss, and Gergely Sipos. Solving the Grid Interoperability Problem by P-GRADE Portal at Workflow Level. In *Grid-Enabling Legacy Applications and Supporting End Users* at HPDC-15, Paris, France, IEEE, June 2006.

[bK+07]   Miachael Klemm, Matthias Bezold, Ronald Veldema, and Michael Philippsen. Reparallelization and Migration of OpenMP Programs. In *Int. Symposium on Cluster Computing and the Grid, Rio de Janeiro, Brazil*, pages 529–540, IEEE, May 2007.

[bKS02]   Herbert Kuchen and Jörg Striegnitz. Higher-Order Functions and Partial Applications for a C++ Skeleton Library. In *Proceedings of the ISCOPE02*. ACM, Nov. 2002.

[bL+01]   Gregor von Laszewski, Ian Foster, Jarek Gawor, and Peter Lane. A Java Commodity Grid Kit. *Concurrency and Computation: Practice and Experience*, 13(8–9):643–662, July/Aug. 2001.

[bL+04]   Sang Lim, Geoffrey Fox, Shrideep Pallickara, and Marlon E. Pierce. Web Service Robust GridFTP. In *Parallel and Distributed Processing Techniques and Applications*, pages 725–730, June 2004.

[bL+06]   Jan Dünnweber, Sergei Gorlatch, Martin Griebl, Eduardo Argollo, and Christian Lengauer. Making a Task Farm Component Parallelize Loops for the Grid. In Thierry Priol Sergei Gorlatch, Marian Bubak, editor, *Proceedings of the CoreGRID Integration Workshop*, pages 93–104. CYFRONET Poland, Oct. 2006.

[bL+07]   Philipp Lüdeking, Jan Dünnweber, and Sergei Gorlatch. A Software Component for Efficient Genome Processing on the Grid. In *Making Grids Work*, Springer, June 2008.

[bLa74]   Leslie Lamport. The Parallel Execution of do Loops. In *Communications of the ACM*, volume 17, 2, pages 83–93. ACM Press, Feb. 1974.

[bLC87]   William E. Lorensen and Harvey E. Cline. Marching Cubes: A High Resolution 3D Surface Construction Algorithm. *Computer Graphics*, 21(4):163–169, July 1987.

[bLe66]   Vladimir I. Levenshtein. Binary Codes Capable of Correcting Insertions and Reversals. *Soviet Physics Dokl.*, 10:707–710, Feb. 1966.

[bLe93]   Christian Lengauer. Loop Parallelization in the Polytope Model. In *International Conference on Concurrency Theory*, pages 398–416, Aug. 1993.

[bLo07]   Steve Lohr. I.B.M. to Push 'Cloud Computing', Using Data From Afar. In *New York Times*, Nov. 15th 2007.

[bM+02]   Ravi K. Madduri, Cynthia S. Hood, and William E. Allcock. Reliable File Transfer in Grid Environments. In *Proceedings of the 27th IEEE Conference on Local Computer Networks*, pages 737–738, Washington, USA, IEEE, Nov. 2002.

[bM+06]   Jens Müller, Martin Alt, Jan Dünnweber, and Sergei Gorlatch. Clayworks: A System for Collaborative Real-Time Modeling and High-Performance Simulation. In *2nd IEEE International Conference on e-Science and Grid Computing*, Amsterdam, Dec. 2006.

[bME05]   Hashim H. Mohamed and Dick H.J. Epema. The Design and Implementation of the KOALA Co-Allocating Grid Scheduler. In *Proceedings of the European Grid Conference*, pages 640–650, Amsterdam, Feb. 2005.

[bNW70]   Saul B. Needleman and Christian D. Wunsch. A General Method Applicable to the Search for Similarities in the Amino Acid Sequences of Two Proteins. *Journal of Molecular Biology*, 48:443–453, Mar. 1970.

[bP+98]   František Plášil, Dušan Bálek, and Radovan Janeček. SOFA/DCUP: Architecture for Component Trading and Dynamic Updating In *Configurable Distributed Systems* (4th ICCDS), Maryland, USA, IEEE, Jun. 1998

[bP+02]   Larry Peterson, Tom Anderson, David Culler, and Timothy Roscoe. A Blueprint for Introducing Disruptive Technology into the Internet. In *Proceedings of the First ACM Workshop on Hot Topics in Networking (HotNets)*, Oct. 2002.

[bP+07]   Arumugam Paventhan, Kenji Takeda, Simon J. Cox, and Denis A. Nicole. MyCoG.NET: A Multi-Language CoG Toolkit. *Concurrency and Computation – Practice and Experience*, 19(14):1885–1900, Sept. 2007.

[bPK08]   Michael Poldner and Herbert Kuchen. On Implementing the Farm Skeleton. In *Parallel Processing Letters*, Vol. 18, Mar. 2008.

[bR+01]   Thomas Rauber, Robert Reilein-Ruß, and Gudula Rünger. ORT – A Communication Library for Orthogonal Processor Groups. In *Proceedings of the ACM/IEEE Supercomputing Conf. 2001 (SC'01)*, Denver, Colorado, USA, Nov. 2001.

[bR+05]   Kees van Reeuwijk, Rob V. van Niewpoort, and Henri E. Bal. Developing Java Grid Applications with Ibis. In *Proceedings of the 11th International Euro-Par Conference*, pages 411–420, Lisbon, Portugal, Sept. 2005.

[bR+06]   Ioan Raicu, Ian Foster, Alex Szalay, and Gabriela Turcu. AstroPortal: A Science Gateway for Large-scale Astronomy Data Analysis. In *Proceedings of the TeraGrid Conference*, Indianapolis, USA, June 2006.

[bS+02]   Walter Binder, Giovanna Di Marzo Serugendo, and Jarle Hulaas. Towards a Secure and Efficient Model for Grid Computing Using Mobile Code. In *8th ECOOP Workshop on Mobile Object Systems: Agent Application and New Frontiers, Malaga, Spain*, June 2002.

[bS+04]   Martin Schmollinger, Kay Nieselt, Michael Kaufmann, and Burkhard Morgenstern. Dialign p: Fast Pair-Wise and Multiple Sequence Alignment Using Parallel Processors. In *BMC Bioinformatics 5*. BioMed Central, Sept. 2004.

[bS+05]   Keith Seymour, Asim YarKhan, Sudesh Agrawal, and Jack Dongarra. Netsolve: Grid Enabling Scientific Computing Environments. In L. Grandinetti, editor, *Grid Computing and New Frontiers of High Performance Processing*, Advances in Parallel Computing, pages 33–51. Elsevier, Nov. 2005.

[bG+09]   Sergei Gorlatch, Jens Müller, Martin Alt, Jan Dünnweber, Hamido Fujita, and Yutaka Funyu. Clayworks: Toward User-Oriented Software for Collaborative Modelling and Simulation. In *Journal on Knowledge-Based Systems*, Elsevier, Mar. 2009.

[bSW81]   Temple F. Smith and Michael S. Waterman. Identification of Common Molecular Subsequences. *Journal of Molecular Biology*, 147:195–197, Mar. 1981.

[bSw96]   Wim Sweldens. The Lifting Scheme: A Custom-Design Construction of Biorthogonal Wavelets. *Applied and Computational Harmonic Analysis*, 3(2):186–200, Apr. 1996.

[bT+00]   Bryan Talbot, Shu-Jia Zhou, and Glenn Higgins. Review of the Cactus Framework. In *Third Round of Scientific Grand Challenge Investigations*, Oct. 2000.

[bT⁺08]   Johannes Tomasoni, Jan Dünnweber, Sergei Gorlatch, Michael Classen, Philipp
           Classen, and Christian Lengauer. LooPo-HOC: A Grid Component with Embed-
           ded Loop Parallelization. In *Grid Computing: Achievements and Prospects*, Springer,
           Aug. 2008.

[bV⁺02]   Francisco Valera, Aymeric de Solages, Enrique Vázquez, and Luis Bellido. Paral-
           lelism and Messaging Services in a J2EE-based E-Commerce Brokering Platform. In
           5th International Conference on Electronic Commerce Research. ACM Press, Oct.
           2002.

[bVa90]   Vaidy Sunderam. PVM: A Framework for Parallel Distributed Computing. In *Con-
           currency: Practice and Experience 2*, pages 712–721, Dec. 1990.

[bW⁺03]   Von Welch, Frank Siebenlist, Ian Foster, John Bresnahan, Karl Czajkowski, Jarek
           Gawor, Carl Kesselman, Sam Meder, Laura Pearlman, and Steven Tuecke. Security
           for Grid Services. In *International Symposium on High Performance Distributed
           Computing*, HPDC-12, IEEE, Washington, USA, June 2003.

[bW⁺05]   January 3rd Weiner, Geraint Thomas, and Erich Bornberg-Bauer. Rapid Motif-
           Based Prediction of Circular Permutations in Multi-Domain Proteins. *Bioinformatics*,
           21:932–937, Apr. 2005.

[bW⁺06]   James A. Kohl, Torsten Wilde, and David E. Bernholdt. Cumulvs: Interacting
           with High-Performance Scientific Simulations, for Visualization, Steering and Fault
           Tolerance. *International Journal of High Performance Computing Applications*,
           20(2):255–285, May 2006.

[bWL07]   Zhiang Wu and Junzhou Luo. Dynamic Multi-Resource Advance Reservation in Grid
           Environments. In *Network and Parallel Computing*, China, pages 13–22, Sept. 2007.

[bWo86]   Michael Wolfe. Loop Skewing: The Wavefront Method Revisited. In *Journal of
           Parallel Programming*, 15(4):279–293, Oct. 1986.

[bY⁺03]   Lingyun Yang, Jennifer Schopf, and Ian Foster. Conservative Scheduling: Using Pre-
           dicted Variance to Improve Scheduling Decisions in Dynamic Environments. In *Su-
           perComputing (SC'03)*, Nov. 2003.

(c) Online References

[cAl08]   Argonne National Laboratory. The Message Passing Interface (MPI), 2008.
           http://www-unix.mcs.anl.gov/mpi

[cAM07]   Ahmed Moustafa. The JAligner Library for Biological Sequence Alignment, 2007.
           http://jaligner.sourceforge.net

[cAo96]   Eecs Winter Namitha.
           Apache Server, 1996. citeseer.ist.psu.edu/352134.html

[cAp03a]   The Apache SOAP Web Site, 2003. http://ws.apache.org/soap

[cAp03b]   Apache Organization. The Apache Web Services Project: Axis, 2003.
           http://ws.apache.org/axis

[cAp03c]   Apache Organization. The Apache Struts Framework for Java Web applications, 2003.
           http://struts.apache.org

[cAp04]   Apache Organization. Jakarta BSF: The Bean Scripting Framework, 2004.
           http://jakarta.apache.org/bsf

[cAp06]   Apache Organization. Apache WS Addressing, 2006.
           http://ws.apache.org/ws-fx/addressing

[cB⁺02]   Henry Bal, Pieter Adriaans, and Bob Hertzberger. Virtual Laboratory for e-Science,
           2002. http://www.vl-e.nl

[cBl06]   Paul E. Black. "octree". in *Dictionary of Algorithms and Data Structures* [online],
           Paul E. Black, ed., U.S. National Institute of Standards and Technology, 3rd January
           2006.
           http://www.nist.gov/dads/HTML/octree.html

[cCB07]   Murray I. Cole and Anne Benoit. eSkel: the Edinburgh Skeleton library, 2007.
           http://homepages.inf.ed.ac.uk/mic/eSkel
[cCC05]   CCA Forum. CCA Glossary, 2005. http://www.cca-forum.org/glossary
[cCh06]   Aj Chen. The Web2x Semantic Publishing Tool, 2006.
           http://www.web2express.org/search/search.html
[cCN05]   CoreGRID Network of Excellence. Basic Features of the Grid Component Model
           (GCM). Technical Report D.PM.04, CoreGRID Institute on Component-based Pro-
           gramming, 2005. http://coregrid.net/mambo/images/stories/Deliverables/d.pm.04.pdf
[cCO04]   CORBA/IIOP v3.0.3. Object Management Group, 2004.
           http://www.omg.org/technology/documents
[cCO07]   Object Management Group.
           The CORBA Component Model, 2007.
           http://www.omg.org/technology/documents/formal/components.htm
[cD$^+$06] Jan Dünnweber, Philipp Lüdeking, Cătălin L. Dumitrescu, Eduardo Argollo, and
           Sergei Gorlatch. The HOC-SA Globus Incubator Project.
           http://dev.globus.org/incubator/hoc-sa
[cDu06]   Dutch University Backbone. The Distributed ASCI Supercomputer 2 (DAS-2), 2006.
           http://www.cs.vu.nl/das2
[cEc03]   Eclipse Foundation. Eclipse – An Open Development Platform, 2003.
           http://www.eclipse.org
[cEr06]   Thomas Erl. Serviceorientation – About the Principles, 2006
           serviceorientation.org
[cFe08]   Scott Ferguson. IBM Investing $360M in Cloud Computing Center, In *IT Infrastruc-
           ture – eWeek*.
           http://www.eweek.com, Aug. 1st 2008.
[cFr99]   The Fractal Web site, 1999. http://fractal.objectweb.org
[cGC05]   Free Software Foundation. GNU C Compiler GCC Manual, 2005.
           http://gcc.gnu.org/onlinedocs/gcc-4.0.0/gcc
[cGF04]   GridForge Working Group. Global Grid Forum (GGF), 2004.
           http://www.ggf.org/ogsa-wg
[cGl96]   Globus Alliance, 1996.
           http://www.globus.org
[cGl05]   Globus Team. The Dynamically-Updated Request Online Coallocator DUROC, 2005.
           http://www.globus.org/duroc
[cGM05]   Google Inc. Google Maps, 2005. An Example Web 2.0 Application
           http://maps.google.com
[cGr08]   Galen Gruman. What Cloud Computing Really Means. In *InfoWorld Online*, in-
           foworld.com, Apr. 2008.
[cGS03]   GNU Scientific Library. Network Theory Ltd., 2003.
           http://www.gnu.org/software/gsl
[cHC07]   CERN European Organization for Nuclear Research. LHC – the Large Hadron Col-
           lider, 2007. http://lhc.web.cern.ch/lhc
[cIB03a]   The IBM Websphere Application Server, 2003.
           http://www-306.ibm.com/software/webservers/appserv/was
[cIB03b]   Jeffrey Liu and Yen Lu.
           Building Interoperable Web Services, 2003.
           http://www-128.ibm.com/developerworks/ webservices/library/ws-jsrart
[cIB07a]   IBM Corporation. Grid Computing on Developerworks Online
           www.ibm.com/developerworks/grid
[cIB07b]   IBM Research. Hyper/J:
           Multi-Dimensional Separation of Concerns, 2007.
           http://www.research.ibm.com/hyperspace/MDSOC.htm
[cJB07]   JBoss Network. Hibernate – Persistence for Idiomatic Java, 2007.
           http://www.hibernate.org

[cMD08]   Mozilla Developer Center.
          Asynchronous JavaScript and XML (AJAX), 2008.
          http://developer.mozilla.org/en/docs/AJAX
[cOA04]   OASIS Technical Committee.
          WSRF: The Web Service Resource Framework, 2006.
          http://www.oasis-open.org/committees/wsrf
[cOA06]   OASIS Technical Commettee.
          Web Services Notifications (WS-N), 2006.
          http://www.oasis-open.org/committees/wsn
[cOR05]   Tim O'Reilly.
          What is Web 2.0?, 2005.
          http://www.oreillynet.com/pub/a/oreilly/tim/news/2005/09/30/what-is-web-20.html
[cOSO08] OSOA Community. Service Component Architecture (SCA) Specifications, 2008.
          http://www.osoa.org
[cPA07]   INRIA.
          The ProActive Web site, 2007. http://www-sop.inria.fr/oasis/ProActive
[cRe79]   Trygve Reenskaug. Models - Views - Controllers.
          Technical Note, Xerox Parc, 1979. http://folk.uio.no/trygver
[cRR07]   David Heinemeier Hansson. Ruby on Rails, 2007. http://www.rubyonrails.org
[cSC05]   Borja Sotomayor and Lisa Childers. *Globus Toolkit 4: Programming Java Services*.
          (book edition: Morgan Kaufmann), 2005. online tutorial:
          http://gdp.globus.org/gt4-tutorial
[cSi06]   Singular Systems Software. JEP: The Java Math Expression Parser, 2006.
          http://www.singularsys.com/jep
[cSM97]   Sun Microsystems. The JavaBeans API Specification, 1997.
          http://java.sun.com/products/javabeans/docs/spec.html
[cSM99]   James Duncan Davidson and Danny Coward.
          The Java Servlet Specification, 1999.
          citeseer.ist.psu.edu/davidson99java.html
[cSM04]   Sun Microsystems. The Java Architecture for XML Binding, 2004.
          http://jaxb.dev.java.net
[cSM05a] Sun Microsystems.
          Java Server Faces, 2005. http://java.sun.com/javaee/javaserverfaces
[cSM05b] Sun Microsystems.
          Java Web Start, 2005. http://java.sun.com/products/javawebstart
[cSM05c] Sun Microsystems.
          JXTA Technology, 2005. http://www.sun.com/software/jxta
[cSM06a] Sun Microsystems. Java RMI over IIOP Protocol Specification, 2006.
          http://java.sun.com/products/rmi-iiop
[cSM06b] Sun Microsystems. Java Database Connectivity (JDBC API), 2006.
          http://java.sun.com/technologies/database
[cSM07a] Sun Microsystems. Java Distributed Systems Home Page, 2007.
          http://www.sun.com/rmi
[cSM07b] Sun Microsystems.
          Java Message Service JMS, 2007. http://java.sun.com/products/jms
[cSM07c] Sun Microsystems.
          Java Standard Edition, 2007. http://java.sun.com/javase
[cSM07d] Sun Microsystems.
          Jini Specification, 2007. http://java.sun.com/jini
[cSM07e] Sun Microsystems.
          The JavaSpaces Specification, 2007. http://java.sun.com/products/javaspaces
[cSP08]   Spring Source Global, Inc. The Spring Framework, 2008.
          http://www.springframework.org
[cSo07]   Southampton University.
          The Semantic Grid, 2007. http://www.semanticgrid.org

[cSQ99]   ANSI/ISO/EIC
            International Standard. Database Language SQL, 1999.
            http://www.techstreet.com/features/ISO_IEC_9075.html
[cTay02]   John Taylor.
            e-Science Definition, 2002. http://www.e-science.clrc. ac.uk
[cTD07]   PDS group TU-Delft 2005.
            GRun: A Tool To Start Jobs on a Globus-based Grid.
            http://www.st.ewi.tudelft.nl/koala/grun_doc.txt
[cTo06]   Tokyo Institute of Technology. The TSUBAME Grid Cluster, 2006.
            http://www.gsic.titech.ac.jp
[cTT04]   Sandholm Thomas, Steve Tuecke, Jarek Gawor, and Rob Seed. Java OGSI Hosting
            Environment Design – A Portable Grid Service Container Framework. In *Globus on-
            line documentation*. Globus Alliance, 2004. http://dsd.lbl.gov/SGT/GlobusDocs
[cUC08]   Unicore Forum e.V.
            UNICORE-Grid 2008 http://www.unicore.org
[cUD07]   OGSA-DAI Project Team. The Open Grid Service Architecture - Data Access and
            Integration OGSA-DAI, 2007. http://www.ogsadai.org.uk
[cUP97]   University of Passau. The Polyhedral Loop Parallelizer: LooPo, 1997.
            http://www.infosun.fim.uni-passau.de/cl/loopo
[cW396]   W3C Recommendation.
            eXtensible Markup Language XML, 1996. http://www.w3.org/XML
[cW399]   James Clark and Steve DeRose.
            XPath: The XML Path Language, 1999. http://www.w3.org/TR/xpath
[cW302]   W3C Recommendation.
            Web Services, XML Protocol Recommendations, 2002.
            http://www.w3.org/2002/ws
[cW304]   W3C Recommendation.
            The Resource Description Framework, 2004. http://www.w3.org/RDF
[cW306]   W3C Recommendation.
            The Web Services Description Language, 2006.
            http://www.w3.org/TR/wsdl20-primer
[cW307]   Don Chamberlin, Peter Fankhauser, Massimo Marchiori, and Jonathan Robie.
            XQuery: An XML Query Language, 2007. http://www.w3.org/TR/xquery
[cW+03]   January 3rd Weiner, Geraint Thomas, and Erich Bornberg-Bauer.
            Raspodom Results, 2003. http://www.uni-muenster.de
            Evolution.ebb/Services/storage/raspodom
[cWi08]   John Willis. What's In A Name? Utility vs. Cloud vs. Grid. In *Data Knowledge Center
            Online*, dataknowledgecenter,com, Mar. 25th 2008
[cYD08]   Yahoo Developer Network Yahoo! Launches World's Largest Hadoop Production Ap-
            plication, 2008. http://developer.yahoo.com/blogs/hadoop/2008/02

Index